Rockville Bridge

Rails Across the Susquehanna

Grif Teller, collection of Ken Murry

Dan Cupper

To Jane—
Dan Cupper
18 DEC 2004

WITHERS PUBLISHING

528 Dunkel School Road, Halifax, Pennsylvania 17032

Table of contents

Cover: *Along the Susquehanna*, watercolor, 22" x 30", ©2002 Ted Rose
Rear cover: Rodney DiPaolo

Layout, design, and typesetting by
Withers Publishing, Halifax, Pennsylvania
Printed and bound by Paulhamus Litho, Montoursville, Pennsylvania
Manufactured entirely in the United States of America

First Printing - March 2002
Second Printed - July 2002

International Standard Book Number: 1-881411-34-6

Introduction

The Susquehanna River and the islands, towns, and cities that fall along its 440-mile-long journey to the Chesapeake Bay have inspired magazine articles and even entire coffee-table books. Unlike many other waterways in Pennsylvania, it cuts directly across the ridge-and-valley topography created by the Allegheny Mountains, rather than running parallel to them.

From an engineering point of view, the ease or difficulty of building a bridge across the Susquehanna depends on the location. For much of its length, the river measures a half-mile to three-quarters of a mile wide. Some portions are deep, but many passages around Harrisburg, Pennsylvania's capital city, are unusually shallow, measuring just two to three feet deep in places. During times of low water, it is possible to wade across from shore to shore, and during the 1930s, one man from nearby Marysville, Pa., even drove a car across.

The spot where the river pierces Blue Mountain, five miles north of Harrisburg, has been the site of three railroad bridges, all of them connecting eastern and western Pennsylvania and, at the same time, linking East Coast with Midwest. While it was still under construction, the third and largest Rockville Bridge – the present 48-arch, stone-and-concrete structure, containing 220,000 tons of white sandstone – was being hailed as a landmark both on the Pennsylvania Railroad and in engineering circles worldwide.

Although it has suffered damage from the elements of nature and has been patched and strengthened, the bridge continues to withstand the ravages of flooding and time. After serving for a century under four owners, it remains a lasting monument to visionary planners, skillful builders, and generations of travelers, shippers, and railroaders. ★

A feather of white smoke marks the wintry passage of an an eastbound passenger train about 1938.
Grif Teller, collection of Ken Murry

RAILROAD BRIDGE ACROSS THE SUSQUEHANNA, NEAR HARRISBURG, PENNSYLVANIA.

above, ***A view looking west shows the crossing of the Pennsylvania Canal after PRR replaced its original wooden truss bridge over the waterway with a two-span plate girder bridge. Installation of the iron spans required a masonry center pier in the canal, which the railroad installed in the winter of 1859-60.*** Collection of Ted Xaras

left, ***A news periodical named*** **Gleason's Pictorial,** ***printed in Boston, featured an illustration of the bridge and a short story about it in its edition of June 4, 1853.*** Collection of Rodney DiPaolo

A different illustration was used on Pennsylvania Railroad letterhead, and on currency that was issued by banks and at least one insurance company. Before the federal government printed paper money, state-chartered banks did so, often looking for attractive illustrations with which to decorate their currency engravings. In circulation between 1852 and 1865, bills bearing the Rockville Bridge image in denominations ranging from $3 to $50 were issued by banks in Alabama, Maine, Michigan, New York, South Carolina, and Vermont – but apparently not by any in Pennsylvania. Letterhead, Pennsylvania State Archives; currency, collection of Rodney DiPaolo

1. Slow start on the first bridge

The original single-track bridge over the Susquehanna River at Rockville was the largest and most costly structure on the Pennsylvania Railroad's approximately 250-mile route between Pennsylvania's capital city of Harrisburg and the important industrial center of Pittsburgh. As it turned out, the progress of its construction was anything but smooth.

When the railroad received its charter on April 13, 1846, officials set about raising financing to build a line to reach Pittsburgh from the western terminus of an existing railroad connection from Philadelphia. This link consisted of the state-owned Philadelphia & Columbia Railroad (Philadelphia-Lancaster) and the Harrisburg, Portsmouth, Mount Joy & Lancaster Railroad, both of which PRR later controlled.

Public opinion had been mounting for an all-rail route to provide better service than that offered by the state's own canal-and-rail system, opened in 1834. Named the Main Line of Public Works, that system required 3-1/2 days to travel between Philadelphia and Pittsburgh, using two railroads and two long canal sections.

Backers of the proposed all-rail line had no time to waste. Pressure came quickly from western Pennsylvanians who wanted a railroad to the east regardless of which port city it led to, Philadelphia or Baltimore. As a result, the state Legislature passed an act allowing a rival, the Baltimore & Ohio Railroad, to enter Pittsburgh unless PRR raised $1 million in cash toward stock subscriptions and placed at least 15 miles of line at each end of the route under contract by July 30, 1847. The company hired its chief engineer, John Edgar Thomson, in April 1847, and his immediate task was to locate the right-of-way – within three months.

Opening the bids

On July 15, 1847, some 200 contractors made their way to Harrisburg for the formal opening of bids on construction of about 20 miles of line from Harrisburg to near Aqueduct, an aptly named point northwest of Duncannon where the state's canal crossed the Juniata River. Contractors were required to start work in 30 days and finish in a year. Fourteen firms or individuals were successful in bidding for the contracts covering work on 18 sections, four contractors holding more than one section.[1] A similar letting occurred a week later at Pittsburgh for the western portion, allowing the railroad to meet its legislative deadline by just eight days. A subsequent group of awards covering another 40 miles of track was made at Lewistown, and from that point on thousands of laborers went to work on the line with the most primitive of tools: picks, axes, shovels, carts, and dynamite.

The section that consisted of the bridge was awarded to a firm named Barron, Hague & Tustin, or simply Barron & Co.[2] Special challenges in building such a large structure were apparent from the start. Thomson, chief engineer and later president of the railroad, wrote in October of that year that work had begun on the eastern section of line, report-

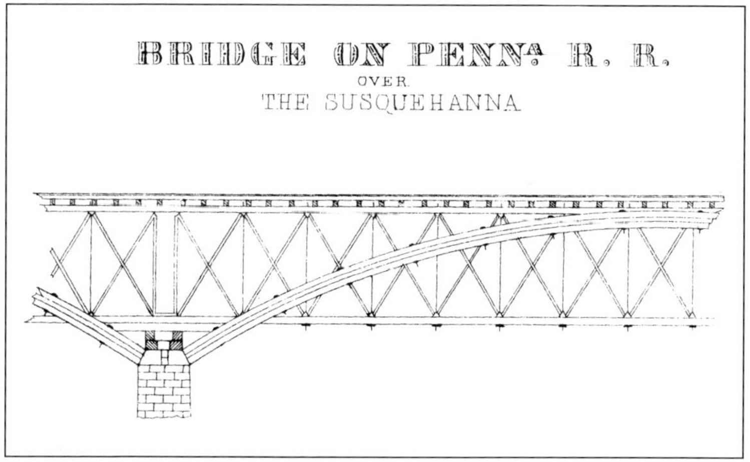

Elevation drawing from General Theory of Bridge Construction, by Herman Haupt, 1853. Collection of Rodney DiPaolo

ing that "it is believed that all, except that including the Susquehanna Bridge, will be finished by the time allotted. The latter is now progressing with much vigor."[3]

The single-track bridge design consisted of 23 wooden 160-foot-long Howe truss spans, measuring 18 feet deep and reinforced with wooden arches. They rested on 22 stone piers six feet wide at the top, and two abutments. In all, the bridge measured 3,680 feet long.[4] Most of the rest of the more than 100 bridges on PRR's 133-mile-long eastern division, from Harrisburg to Altoona, Pa., at the base of the Alleghenies, were also built as Howe wooden trusses. As a result, they resembled the Rockville Bridge in most ways except length. Most of the 73 bridges between Altoona and Pittsburgh, about 117 miles, were built of stone or iron.[5]

above, ***The stone masonry used in the west abutment is prominent in this stereoview looking east toward Rockville.***
Collection of author

Trouble, and more of it

Barron & Co. ran into difficulty in completing its work, and abandoned the contract. A young civil engineer employed by PRR, Herman Haupt, who later became its general superintendent and chief engineer, described the construction, cryptically noting that an interruption of work took place. He wrote: "The foundations of the 22 piers were commenced and carried above water in 16 weeks; but after the eighth pier was founded, operations were suspended for a period of two months not included in this statement."

right, ***In another view looking east, a watchman's shanty guards the western approach to the single-track bridge.***
Pennsylvania State Archives, MG-199

The masonry portion of the contract was re-awarded to Holman & Simons of Harrisburg (also recorded as Holman, Simon & Burke). Robert McAllister of Juniata County assisted that firm, boating some of the cut and dressed stone 50 miles to the site. Bridge contractor Daniel Stone, who had recently completed a bridge over the Schuylkill at Market Street, Philadelphia, went to work on the truss spans and other elements of the superstructure.[6]

By the time the railroad issued its second annual report on Oct. 31, 1848, Thomson was able to write: "The superstructure of the Susquehanna Bridge . . . is more than one-third raised, and will all be done early the next spring." The masonry was completed in December, and Thomson looked forward to opening the first stretch of the main line, Harrisburg to Lewistown, about 60 miles, in the spring of 1849.[7]

But weather disrupted that timetable. On March 28, 1849, a hurricane – some reports called it a tornado – approaching biblical proportions hit the Harrisburg area, destroying the uncompleted portion of the bridge and damaging part of the completed section. The 14th of the 23 spans had been raised when the storm hit.[8]

The *Democratic Union* daily newspaper reported: "In the afternoon a sudden whirlwind caught the new railroad bridge … and lifted six spans of it from the piers and hurled them into the Susquehanna. The c[h]ords and part of the timbers of three additional spans were piled on the spans blown away, at the time of the accident. The parts of the bridge blown down lodged on the islands in the neighborhood of (Harrisburg) and will nearly all be recovered." On the spans that were damaged, the newspaper reported, the curved arches had not yet been installed and the span timber had not been secured to the piers. By contrast, the spans where those two procedures had been completed remained untouched.[9] Another report claimed that the wind was so powerful as to whip the water 30 feet above river level and carry off 40-foot-long timbers.

Haupt, who became the railroad's chief engineer from 1853 to 1856, later wrote that he was an eyewitness to the storm and its devastation on the bridge, which left a deep impression on him. When he published a book on bridge engineering in 1853, he devoted 28 pages to Rockville Bridge, including minutely detailed calculations about the theoretical stresses, both normal and extraordinary (including wind) on the structure. By contrast, he gave much less space to two other nearby bridges over the Susquehanna – five pages to the Cumberland Valley Railroad bridge at Harrisburg, and a scant two pages to the Clarks Ferry Bridge at Duncannon.

Like the newspaper, he concluded that the unfinished nature of the six spans had led to their collapse: "The contractor was engaged at the time in putting in the arches, and as the diagonal braces could not be permanently introduced until after the arches were in place, he had omitted them, except over the piers and in the middle of the spans. The direction of the storm was nearly at right angles to the bridge . . . The only arrangement that could have secured the bridge in so violent a tornado, would have been a complete system of diagonal bracing, but the accident occurred before these could be introduced ...."[10]

Ready for business

The contractor moved quickly to recover, and the bridge was opened within a few months. It had cost $169,956,[11] of which $96,355 went for masonry. Haupt wrote: "... the whole of the work on the piers and abutments, including an arcade of three full [stone] center arches of 25 feet span, constituting the eastern approach of the viaduct, was completed in one season." The wooden spans were fabricated mostly of white pine, with some members made of white oak.[12]

Near the end of July, the railroad ran a train as far as Millerstown, 30 miles west of Harrisburg and about halfway to the initial goal of Lewistown, for the first time, using the occasion to entertain dignitaries. Aboard were Governor William F. Johnston, the state treasurer of Pennsylvania, and other heads of departments of state government.[13]

On August 23, PRR operated what was apparently the first train to run from Harrisburg to Lewistown, a lumber train. One week later, on the 30th, another train of dignitaries rode from Harrisburg to Lewistown. The next day, on September 1, regular scheduled passenger and freight service began. At first, freight trains ran twice a week, and then three times a week, leaving for the east at 7 a.m. A single passenger train left Lewistown daily at 9:45 a.m. The fare to Harrisburg was $2.10, and to Philadelphia, $5.50.[14]

Although it was now open for business, PRR continued to push construction westward, finally linking with the state's Allegheny Portage Railroad at Duncansville, south of Altoona, in the fall of 1850. Service gradually increased so that by 1853, the line was seeing four trains a day each way—a fast passenger, a slow passenger, a freight train, and a freight and emigrant train.[15] For three years, PRR used the APRR as an interim bridge over the mountains to connect with its western section, but this was unsatisfactory. The state railroad didn't run at night, and it closed each winter when the canals froze over. PRR completed its own line between Johnstown and Altoona on February 15, 1854, filling in the missing link. With this, the railroad cut travel time between Pittsburgh and Philadelphia to as little as 13 hours.

Also during 1854, the railroad built a new water station near Rockville, and laid a siding there to accommodate the freight cars that were now coming to and from another railroad at a new junction. The Dauphin & Susquehanna Railroad, a predecessor of the

top, ***An eastbound freight train has stopped to pose for a stereoview photo with the engine above the three masonry arches on the eastern side. Timbers for bridge renewal are stacked near the railroad embankment, at the foot of a slide used to place and retrieve them from track level.*** Collection of author

right, ***Some of the original 1850s PRR bridges between Altoona and Pittsburgh were stone-arch spans, including the twin-arch example at Bolivar, west of Johnstown, shown in this W.T. Purviance stereoview.*** Collection of author

Visible in this Frederick Gutekunst stereoview showing the east end of the bridge are PRR's Rockville station and timbers stored for renewing the bridge. The track in the foreground is the western end of the Schuylkill & Susquehanna Railroad, a branch of the Reading Railroad system that stretched more than 50 miles from Auburn in the anthracite coal region to Dauphin and Rockville, where it connected with PRR.
Collection of author

Schuylkill & Susquehanna Railroad and, ultimately, the Philadelphia & Reading Railroad (later the Reading Co.), had built a line westward through Stony Creek Valley to the town of Dauphin, Pa., 2-1/2 miles north of Rockville on the same side of the river. It was built to carry coal from anthracite mines at Cold Spring to the canal at Dauphin, a section that was opened in 1851. Later, D&S/S&S extended further east, reaching Pine Grove and Auburn, which was located on the Reading's main line from Williamsport, Shamokin, and Pottsville, Pa., to Philadelphia. It also extended south from Dauphin to Rockville and a connection with PRR, bringing its length to 50 miles. Reading and PRR entered into a trackage rights agreement, which allowed D&S/S&S trains to run over the PRR tracks into Harrisburg, where the Reading was building its Lebanon Valley line from Philadelphia, Pottstown, Reading, and Lebanon.

Built as a single-track structure, the wooden Rockville Bridge very early became a potential bottleneck when the company realized that it needed to double-track its entire main line to handle the amount of business that was being offered it. During 1854, a second track was laid from Harrisburg to the east end of the bridge, a little more than five miles, and about .6 mile of double track already existed at the west end of the bridge. PRR began to aggressively add a second mainline track wherever money and available labor would allow.

With more traffic rolling over the Pennsylvania Railroad, its physical property, including the bridges, began to need maintenance. The bridge had been in use for more than five years when George R. Mowry, a PRR assistant superintendent, reported on January 1, 1855, that it "will require a new coat of paint as soon as the weather, in the spring, will permit."[16] In the next year's report, he noted: "The Susquehanna Bridge has been repaired, and on all the bridges between the Susquehanna Bridge and Altoona, cross-ties have been renewed."[17]

Completing the system

The speed, ease, and all-weather availability of PRR service quickly made the state's canal-and-rail network obsolete. In 1857, the railroad bought the Main Line of Public Works for $7.5 million. It immediately integrated the Philadelphia-Lancaster-Columbia railroad into its own system, expanding its operations so that it now linked Philadelphia with Pittsburgh via a main line of about 350 miles. PRR soon closed and dismantled the Allegheny Portage Railroad, and set up a subsidiary, the Pennsylvania Canal Co., to operate the severed stretches of canal, for which some local and coal business remained. Bit by bit, PRR cut back the navigable portion of the canals.

Freight and passenger business on PRR began to grow immensely. Freight tonnage rose from 159,000 tons in 1853 to 530,000 tons in 1857. Two years later, with PRR's help, the Pittsburgh, Fort Wayne & Chicago Railway was completed, giving the Pennsylvania an outlet to the Midwest and West and increasing its traffic potential. PRR was also investing in other routes through Ohio and Indiana, and was laying plans to connect to Cincinnati; Cleveland; Indianapolis; Louisville, Ky.; and St. Louis, making it one of the early important trunk lines.

Wooden bridges may have been a marvel of planning and efficiency, but they also required continual maintenance. They also were constantly exposed to the threat of fire ignited by sparks from passing steam locomotives, which were running more frequently with the rising tide of traffic. In 1855, two PRR wooden bridges burned from this cause, one west of Lewistown over the Juniata, and the other farther west, over the Little Juniata River[18] – this despite the fact that the railroad had painted all of its wooden bridges with "fire-proof paint."[19] PRR was already beginning to replace wooden spans with iron bridges built in its shops at Altoona. In 1858, Thomas Seabrook, resident engineer, reported: "The bridges have all undergone a thorough examination, and many of them found to be much decayed. This examination leads to the conclusion that about nine years is as much as can be calculated upon for the duration of an uncovered wooden superstructure, and when the liability to fire and expense of [hiring men to do] watching is taken into consideration, shows most conclusively that economy, as well as safety, is in favor of iron bridges."[20]

He also noted that extensive repairs had been made to Rockville Bridge: "The sheet iron roof, which had become rusty and full of holes, has been removed and replaced by galvanized iron; the track has been relaid with new iron, upon a continuous stringer securely bolted to the under stringers and floor beams; the ice breakers, which had been much injured by the ice freshet last winter, have been thoroughly repaired, and the upper ends of piers secured by rip rap; the tin covering and hand rail at the sides of the bridge have received a coat of brown zinc paint." The roof to which he alludes may refer to sheet-metal coverings, installed with a pitch outward toward the river, that ran along the top chords of the bridge, protecting not only those members but also the arches, or, perhaps more likely, to a metal sheathing that was positioned beneath rail level to keep hot ashes, coals, or cinders off the wooden bracing.[21]

During the following year, 1859, the railroad discovered even more maintenance needs: "The wall plates, bolsters, and end segments of the arches were found to be defective, and have been renewed: arch struts and set screws have been added, also additional diagonal braces, and such other repairs were made as were required to place the bridge in good condition."[22] At the same time, PRR took advantage of the close of navigation at the onset of the winter of 1859-60 to install a masonry center pier in the canal at Rockville to replace the original through-truss bridge there with a two-span iron girder bridge.

Northern Central bridge at Dauphin

Just upstream two miles, another railroad span had been built across the Susquehanna. In 1855, the Northern Central Railway, which was affiliated with and later became part of PRR, awarded a $750,000 contract to the McCollum Bridge Co. to build a 19-span,[23] 3,846-foot-long structure.[24] Its completion in March 1858[25] allowed NCR four months later to open its main line – a north-south route – from Baltimore through York, Bridgeport (Lemoyne), Fairview (Enola) and what is now Marysville, to Millersburg and Sunbury, Pa. There, NCR connected with other PRR-affiliated lines to Erie, Pa.; Buffalo, N.Y.; and Lake Ontario. At Marysville, NCR connected with PRR to allow freight traffic to be exchanged between the two roads. Their main lines also crossed at grade, at the west end of the NCR bridge.

The opening of the Marysville-Dauphin bridge allowed NCR to run through from Baltimore to Sunbury, but because its line ran along the west side of the Susquehanna, its passenger trains had to make a slow detour to reach Harrisburg. This maneuver involved making a back-up movement via the Cumberland Valley Railroad bridge to pick up and discharge patrons bound to or from Harrisburg, then recrossing to Bridgeport (Lemoyne) before heading north or south again. The result was that through passengers crossed the Susquehanna three times.

Passengers endured this tedious procedure while NCR officials grumbled that "a proper connection between [the Cumberland Valley Railroad bridge at Harrisburg], and the right from the Schuylkill & Susquehanna Rail Road, and the Penn-

Bridge Rules on the Pennsylvania Railroad, 1857

After the railroad was in operation, officials drew up highly specific rules for bridge maintenance and operation. The following excerpts from the 1857 PRR Book of Rules illustrate the priorities:

- All Trains must be run slowly over Bridges. Ten miles per hour or six minutes to the mile is the highest rate of speed allowed in passing them.

- Foremen and Watchmen on the Pennsylvania Rail Road are directed to remove from the vicinity of bridges all combustible materials, allowing nothing to remain within one-fourth of a mile, along the track in either direction, that could conveniently be carried and used for incendiary purposes. They will also examine particularly the masonry at the top of the piers and abutments, remove chips and all similar substances therefrom, and as much as possible protect the timber at the tops of the walls from contact with combustible materials. They will also remove from the vicinity of the track all cross-ties, iron rails, or whatever else might be used in forming obstructions on the road. The utmost vigilance must be used to detect and bring to punishment all persons who place obstructions on the track. Liberal rewards will be given for information that will lead to conviction.

BRIDGE WATCHMEN

- Watchmen on Bridges must keep a constant supply of water upon the Bridge at such intervals as may be required by the Supervisor.

- They shall follow every train with a bucket in hand, and extinguish any coals that may have been dropped from the Engine.

- Where the road-way is on the top, they shall pass through and underneath the Bridge at least once every day to examine the timber, and report any failure or decay thereof to the Supervisor.

- They shall prevent all persons not duly authorized, from crossing the Susquehanna Bridge, and such other Bridges as the General Superintendent may direct.

- When required, they shall keep a Time-book, and note the exact time occupied by each train in passing, and the hour of the day at which it passed.

- In winter they shall remove all snow from off the Bridge and track as quickly as possible after it has fallen.

With the town of Dauphin visible beneath the closest span, this view shows the early appearance of the Northern Central Railway bridge over the Susquehanna at Marysville. Pennsylvania State Archives

Rebuilding of the NCR bridge converted much of it into a covered bridge, as seen in this W.T. Purviance stereoview. Collection of author

Early Railroad Bridges on the Lower Susquehanna

When it opened in 1849, the first Rockville Bridge was the third railroad bridge to cross the Susquehanna River.

Location	*Railroad*	*Opening Date of Earliest Bridge or Line Segment*
Perryville, Md.	Philadelphia, Wilmington & Baltimore (PRR)	November 28, 1866
Perryville, Md.	Baltimore & Ohio	July 11, 1886
Columbia-Wrightsville, Pa.	Columbia Bridge Co.	Highway bridge opened 1834; rails added later that year. Burned by townspeople and Union troops on June 28, 1863, to halt a northward invasion of Confederate troops.
Shocks Mills, Pa.	Pennsylvania Railroad	January 1, 1905
Harrisburg, Pa.	Philadelphia, Harrisburg & Pittsburgh (Reading)	1890-91
Harrisburg, Pa.	Cumberland Valley (PRR)	January 16, 1839
Rockville, Pa.	**Pennsylvania Railroad**	**September 1, 1849**
Marysville-Dauphin, Pa.	Northern Central Railway	March 1858 (railroad opened July 1858)
Selinsgrove, Pa.	Sunbury & Lewistown (PRR)	1871

Metal sheathing beneath the rails caught hot cinders dropped by steam locomotives to reduce the possibility that they would set ablaze the wooden truss members below.
Collection of Rodney DiPaolo

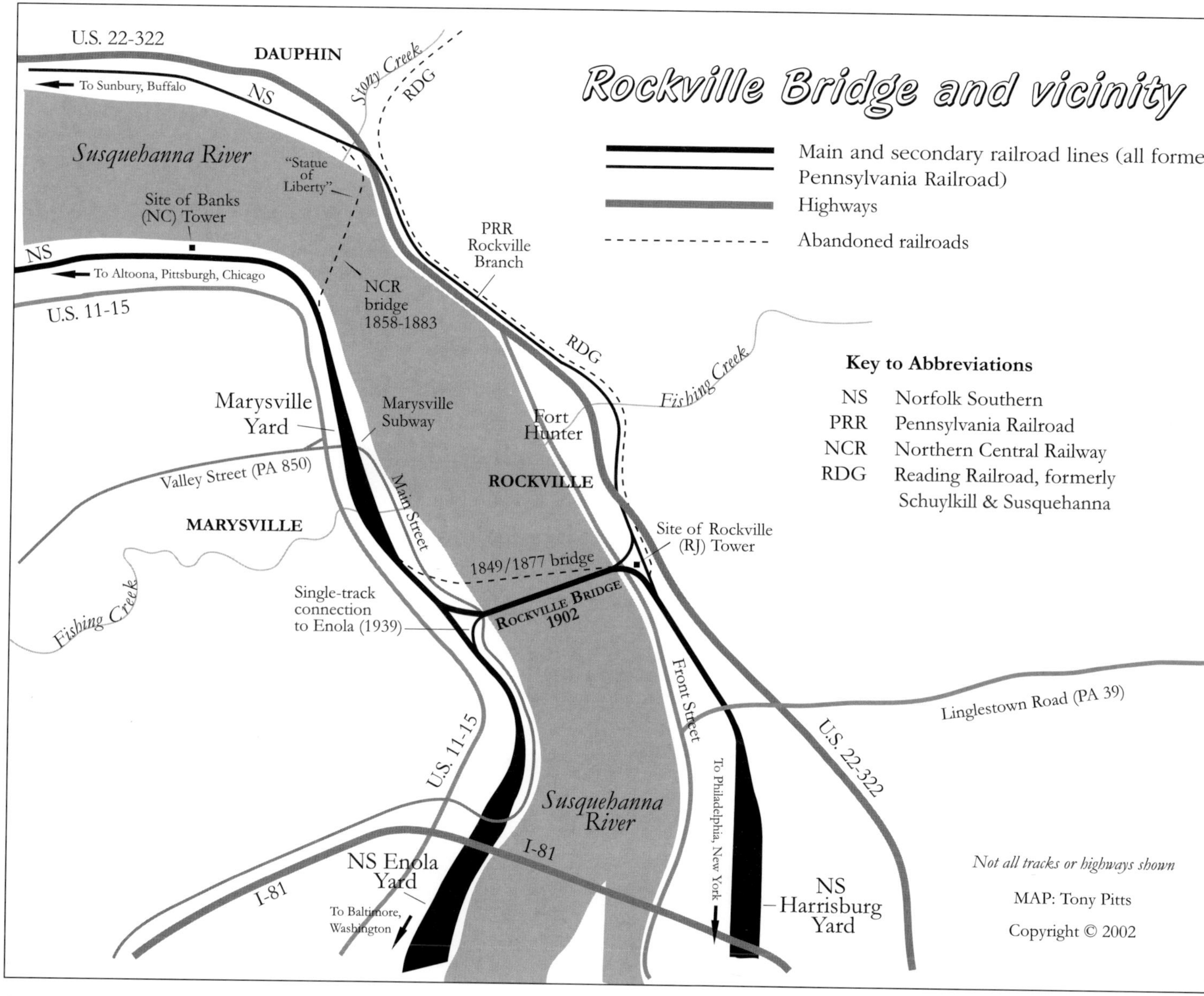

sylvania Rail Road, to use their tracks between Harrisburg and Dauphin, would save nearly an hour in the running time of the road."[26] At some point, the three lines did reach an agreement, because by 1874, NCR trains were crossing the Susquehanna just once – using the CV bridge into Harrisburg, then traveling upriver via PRR to Rockville and S&S to Dauphin, where they re-entered the NCR's own main line to Sunbury and other points north. Later, during a period when relations began to cool between PRR and the S&S's parent, the Reading, PRR built its own branch between Rockville and Dauphin to bypass the S&S segment and allow the retirement of the NCR bridge. But that would not come for another decade. ★

In a winter view, the first Rockville Bridge stretches eastward across a chilly Susquehanna. In the foreground is the track of the Northern Central Railway.
Collection of Walt Ready

2. Troubles at Rockville Bridge

When the Civil War broke out in 1861, railroads became strategically important for the first time in military history. Railroads figured into war in several ways: They were used to move troops and materiel from one theater to another, they functioned as supply lines for forward action during battle, and they were targets for enemy sabotage.

The Pennsylvania Railroad was deeply identified with the Union, and not just because its tracks ran exclusively through a Northern state and connected important Northern cities. Simon Cameron, who was a stockholder and former director of the PRR-affiliated Northern Central Railway, was President Abraham Lincoln's first secretary of war. Early in the war, Thomas A. Scott, a PRR vice president, had been commissioned a colonel in the District of Columbia militia and served as assistant secretary of war under Cameron with responsibility for military railroads and telegraphy. After Cameron resigned under fire, Scott returned to PRR but continued to work closely with military officials to support the Union cause. Also, former PRR chief engineer Brigadier General Herman Haupt, who had been present during the construction of Rockville Bridge, was chief of Construction and Rehabilitation of the United States Military Railroads. In that role, he directed the rebuilding of many sabotaged bridges in short order, and often took direct charge of dispatching the movements of troop, materiel, and hospital trains.

Finally, PRR was a strategically placed line, on several counts. For one thing, it carried munitions and other supplies from the iron and industrial center of Pittsburgh. For another, its location made it less vulnerable than the Baltimore & Ohio Railroad – which was frequently damaged and cut by the Confederates – for moving Union troops and artillery east and west.

Rockville Bridge was directly threatened twice, each time during General Robert E. Lee's invasions of the North. This occurred when his troops entered Maryland in the fall of 1862 during the Antietam campaign, and when they entered Pennsylvania in the summer of 1863, which resulted in the landmark Battle of Gettysburg. In both cases, the occupation of Harrisburg was the objective and the bridge was considered to be a military target to be controlled or destroyed.

The three masonry arches at the east end of the bridge are evident in this Purviance cabinet-card view looking west from high on Blue Mountain.
Collection of author

September 1862

On September 8, 1862, Confederate Major-General John G. Walker rode into Frederick, Md., which the rebels had already occupied. He reported to Lee, who then revealed his overall strategy because Walker's division was a detached one that would likely be ordered from place to place to assist other units as needed. Here is Walker's description of Lee's explanation:

> "In 10 days from now, if the military situation is then what I confidently expect it to be after the capture of Harper's Ferry, I shall concentrate the army at Hagerstown (Maryland), effectually destroy the Baltimore & Ohio (Rail)road, and march to this point," placing his finger at Harrisburg, Pa. "That is the objective point of the campaign.
>
> "You remember, no doubt, the long bridge of the Pennsylvania Railroad over the Susquehanna, a few miles west of Harrisburg. Well, I wish effectually to destroy that bridge, which will disable the Pennsylvania Railroad for a long time. With the Baltimore & Ohio in our possession, and the Pennsylvania Railroad broken up, there will remain to the enemy but one route of communication with the West, and that very circuitous, by way of the Lakes. After that I can turn my attention to Philadelphia, Baltimore, or Washington, as may seem best for our interests."[1]

As events turned out, Lee's armies were turned back, and the invasion of the North was reversed after the Battle of Antietam.

June-July 1863

Lee again threatened the North the following summer, hoping again to take Harrisburg. From there, he could move on Philadelphia or New York, or choke a vital Union supply of coal by seizing the anthracite mines of northeastern Pennsylvania. During mid-June 1863, rebel troops formed a 100-mile-long line that began sweeping up the Cumberland Valley. They soon occupied Chambersburg, York, Wrightsville, and Carlisle, all within striking distance of Harrisburg. Lee had ordered Lieutenant General Richard S. Ewell to take the capital city if possible. One strategy was to cross the Susquehanna at Wrightsville into Columbia, seize Lancaster and attack Harrisburg from the east.

While many Harrisburg residents were complacent, others began to grow nervous. Although PRR's traffic was heavy – the railroad had moved a record 12,000 freight cars between Pittsburgh and Philadelphia from January to June of that year[2] – it ceased operations in light of the impending attack. East-west mail was diverted to New York Central lines via Cleveland. PRR moved its engines and cars elsewhere to safety and turned its Harrisburg roundhouse over for military use as a headquarters for two companies of militia. Commented the Harrisburg *Patriot and Union* newspaper: "The action of the railroad company, in suspending their important operations in the machine shops, is patriotic and commendable."[3] PRR asked permission of the Reading to store its locomotives and cars on lesser-used parts of that railroad for safe keeping, and the Reading cooperated.

Charles E. Smith, president of the Reading, was taking no chances, either, wiring to that line's general superintendent at Harrisburg on June 16: "You had better bring away from Harrisburg all our cars, and be prepared to move all valuables from there. The rebels will be on the west side of the river tomorrow, I think."[4] At the Cumberland Valley Railroad roundhouse at Bridgeport (Lemoyne), the walls were pierced for musketry to defend against a northward rebel advance.

Alarmed, both President Lincoln and Pennsylvania Governor Andrew Curtin issued a call for tens of thousands of volunteers to defend Pennsylvania, but the enlistment term of six months was too long, and the prospect of pay too scant, to attract many men. They reduced the term to three months, or less if the Confederate threat receded,[5] and Pennsylvanians slowly began to sign up. But in the meantime, the New York State National Guard offered as many as 10,000 militiamen who were already well-trained and in service, and the offer was accepted by Lincoln's second secretary of war, Edwin M. Stanton. As a result, it was almost exclusively New Yorkers who took up the defense of Harrisburg and Rockville Bridge. Thirteen regiments of New York soldiers began to pour into Harrisburg by rail; of those, nine were deployed in and around the city, and four were sent to guard Rockville Bridge and the nearby Northern Central Railway bridge connecting Marysville and Dauphin.[6]

Brigadier General Charles Yates, commanding the Second Brigade of the New York State National Guard, arrived at Harrisburg on June 20 and reported to Major-General Darius Couch, commander of the Department of the Susquehanna. The next day, Yates took his Fifth and Twelfth Regiments, comprising about 1,000 men, to Marysville to guard the bridges. They believed that Confederates moving northeastward through the Cumberland Valley might attempt to cross Blue Mountain into Perry County via Sterrett's Gap or another gap, and attack from that direction. Yates reported:

> "Our position, therefore, assumed a very important character, and required very great diligence in checking a movement of that kind. On arriving at [Marysville], the only force we found there consisted of about 50 or 60 men of the Invalid Corps,[7] stationed at block-houses near the bridges. We immediately commenced constructing such earthworks in front and flanks as were deemed necessary, and selected a position to make a determined stand, the left flank of our contemplated line being protected by an obstruction of felled woods, and the narrow pass along the Susquehanna being guarded by such force as could be spared for that purpose. Various detachments from time to time were sent with ax-men, to obstruct and guard the mountain gaps."[8]

During the 17 days that the two regiments were camped at Marysville, the soldiers were "exposed to almost incessant rains, having only their shelter tents to protect them from the inclemency of the whether," Yates wrote. During part of their assignment there, they were augmented by troops of the 13th and 28th New York National Guard, and three companies of the 28th Pennsylvania Militia.

The fears of invasion were real. On June 28, Union cavalry pickets exchanged shots with Confederates along the Carlisle Pike at Oyster Point in present-day Camp Hill, just 2-1/2 miles from downtown Harrisburg. Historians usually consider this to be the high-water mark of Southern incursion. On the night of that same day, Northern troops and townspeople burned the railroad bridge over the Susquehanna at Columbia, 25 miles south of Harrisburg, to block approaching Confederates from crossing. From the 28th through the 30th, rebel forces overran part of the Northern Central Railway, burning 12 bridges on the main line between the Mason-Dixon line and Goldsboro, south of Harrisburg, and 19 more on the branch between York and Wrightsville.

General Couch prepared parts of the Cumberland Valley Railroad and the Camelback (Harrisburg Bridge Co.) bridges over the Susquehanna at Harrisburg for demolition by sawing through part of the westerly span of each in such a way that they could be dropped into the river quickly in the face of an advancing army. Lee's plans to take Harrisburg were interrupted when his troops met the Union Army in the monumental Battle of Gettysburg on July 1-3, after which he began to retreat southward.

By July 7, things began to return to normal, and PRR had resumed operations over its entire route.[9] On that date, Yates and his troops left Marysville as part of a Union movement to pursue the retreating Southern army, but when Lee crossed the Potomac River, the threat was considered to be over. Yates and his men returned to New York City, where they were mustered out later that month.[10]

People standing on the east bank of the Susquehanna give a sense of the size and proportions of the wooden bridge and its stone approach arches. The struts projecting from the bridge railing carry telegraph wires.
Collection of Rodney DiPaolo

Philadelphia stereoview photographer Frederick Gutekunst (1831-1917) set up this view looking west, showing the Pennsylvania Canal and an adjacent residence with a whitewashed fence. While the lounging man in the foreground is unidentified, he may have been either Gutekunst or an assistant. Gutekunst and other photographers often posed people in their photos to add interest or provide scale.
Collection of author

Sparks lead to fire

What General Lee's army could not accomplish, one of the Pennsylvania Railroad's own steam locomotives did, just a few years after the Civil War ended.

Early on the morning of July 17, 1868, sparks from the engine of the *Baltimore Express* passenger train caught fire, and the five westerly spans burned. Rockville citizens, as well as firefighters and a steam fire engine from Harrisburg's Friendship fire company, helped to contain the blaze.[11]

A few days later, a correspondent of the *Harrisburg Telegraph* was en route to Sunbury aboard a train, and noticed that work on a new PRR roundhouse in Harrisburg had come to halt. Finding that everyone had been reassigned to rebuild Rockville Bridge, he filed this report:

> "At Rockville a scene of bustle and activity presented itself beyond description. A regiment of carpenters and laboring men were engaged in framing and conveying timber hither and thither. A construction train was at hand, with cars attached and 'steam up' all the time to carry material to the damaged portion of the structure. On thc Perry County side we observed, also, a large force of workmen employed on shore, on the bridge and in the river. This large force, I am told, is relieved from time to time, both day and night – an excellent arrangement – so as to facilitate the progress of repairing the magnificent structure, and putting it in proper condition for passing trains over at an early day. I noticed that two spans were already on the trestle works, and that men were engaged in putting up the third span of 'false works' or trestles.'"[12]

Using a convenient nearby detour route, PRR barely missed a beat in moving trains. Westbound trains ran from Harrisburg to Rockville as usual, then detoured over the Schuylkill & Susquehanna Railroad to Dauphin, then crossed the Northern Central Railway bridge to Marysville, where they got back on the PRR main line and headed west. Eastbound, the procedure was reversed.

The NCR assigned 10 bridge watchmen, a telegrapher, and a dispatcher to oversee the operation, which raised the traffic levels to between five and eight times the usual number of trains. In the 24-hour period after the fire, NCR handled 49 PRR freights and 15 PRR passenger runs, for a total of 64 PRR trains, in addition to 13 of its own trains, making a grand total of 77 trains, or one about every 20 minutes. The count rose even higher, with a total of 100 trains moved in the period ending at midnight on July 18.[13]

Although the detour was intense, it did not last for long. Repairs were completed in 10 days, and Rockville Bridge was reopened on the evening of July 28, with the *Cincinnati Express* being the first train across about 10 o'clock that night.[14]

PRR picnic grove at Marysville

About the same time, the railroad was beginning to promote local travel on its lines around Harrisburg. In the summer of 1868, PRR built a picnic grove at Marysville and began to sell excursion tickets to Sunday Schools, fire companies, and other groups. Among the facilities were a covered dance pavilion, a bath house, a grille called a barrack range, "where coffee & etc. may be made as wanted," and a cold-water spring. Round-trip fare from Harrisburg was 40 cents for adults, 20 cents for children.

One of the biggest such gatherings took place a month after the bridge was repaired, when local PRR employees and their families held their third annual company picnic, with PRR supplying free transportation. On the morning of August 15, a train consisting of six baggage cars and 20 coaches, carrying about 2,200 passengers, took the first wave of excursionists out of Harrisburg and over the bridge to the grove. The baggage cars were filled with the passengers' tables, chairs, tubs, wagons, picnic baskets, ice-cream freezers, and cookstoves. The engine, No. 35, was decorated with multicolored paper and with the American flag. About noon, a westbound mail train dropped off more picnickers, followed by another special consisting of 17 cars. Throughout the day, dancing, eating, and socializing went on, to the accompaniment of a brass band and a string band. One participant noted that the crowd "cared nothing about reconstruction, rebels, Grant or Colfax. Democrat and Republican met on common ground, and for one day at least forgot business and politics."[15]

At the end of the day, the excursionists packed up and boarded an immensely long train for the six-mile return to Harrisburg. One writer described what he saw this way: "When we were on the bridge with about 10 cars, we could then look back and see the whole train coming slowly around the curve. They were 42 cars, making a number of at least 4,000 persons. Just think of it, 4,000 persons, and nearly one-half of them waving their hats and handkerchiefs out the windows."[16]

Perhaps as a result of the fire repairs in 1868, PRR performed additional work the following year. The annual report for 1869 stated: "The entire floor of the Susquehanna Bridge at Rockville, has been relaid in a new shape, covered with tin and painted; about three-fourths of the floor beams have been renewed, also a number of chord pieces."[17] Earlier, in 1867, the railroad had renewed 700 arch members, 21 diagonal braces, and 18 other parts called wall plates.[18]

But PRR was already well along in its plan to replace its maintenance-intensive and fire-prone wooden bridges with wrought-iron spans. By 1872, engineering consultants later reported, "all wooden bridges in the main line, Harrisburg to Pittsburgh, except the Susquehanna River Bridge at Rockville, had been replaced with permanent structures."[19] By 1875, the railroad counted 173 iron bridges in its main line between Philadelphia and Pittsburgh.[20]

Fire was never far from the minds of railroad engineering officials. In a reverse of the situation that resulted from the 1868 fire on Rockville Bridge, sparks from a freight locomotive on the night of April 3, 1872, ignited a fire on the Northern Central bridge between Marysville and Dauphin. Ten of the 19 spans burned,[21] necessitating another detour while that bridge was rebuilt—NCR trains used the S&S line between Dauphin and Rockville, and the PRR main line between Rockville and Harrisburg.[22]

Single-track bottleneck

Although Rockville Bridge measured less than a mile in length, it still was an operating bottleneck, being the last single-track section of the main line. In 1875, Samuel Vauclain, who later was chairman of the Baldwin Locomotive Works, was an apprentice in PRR's Altoona Shops. At one point, he was assigned to ride an engine on the Middle Division as part of some experimental tests. He wrote: "This one-track wooden structure always yielded to the weight of the locomotive and crossing was sort of an up-and-downhill performance. The speed of all trains when crossing was limited to 4 mph."[23]

As the volume of traffic rose, so did the number of trains, and PRR looked for better, faster, and safer ways of handling them. Technology provided some answers, and the development of the air brake in 1869 allowed trains to run at higher speed with greater control over their ability to stop. PRR was among the first railroads, if not the first, to equip its passenger trains with air brakes.

Another way of improving train handling was the manual block system of signaling. By 1875, the company had adopted this method, and by the following year, when the Altoona-Harrisburg segment was equipped,[24] the railroad was able to announce that its New York-Philadelphia and Philadelphia-Pittsburgh routes were protected in this way.[25] In this system, a line was divided into sections, with 24-hour manned telegraph offices situated every few miles, and only one train allowed in a block at a time. Acting on orders from a centrally located train dispatcher, the block operator who worked in each trackside two-story tower displayed elevated signals to govern the movement of trains; the color indications were red for stop, blue for caution, and white for clear.[26] PRR built a tower at each end of Rockville Bridge, because of the need to smoothly coordinate trains entering and leaving the single-track bridge from the double-track segments on either side of it. Towers were identified by their one- or two-letter telegraphic call symbols (brevity being important in sending messages), and the railroad attempted to link them to some local landmark as a memory aid – the one on the east side was designated RJ, for Rockville Junction, and the one on the west was designated PR, the first two consonants in the word Perry, the name of the county in which it was located. ★

below, ***A view looking east from the west bank of the Susquehanna shows the slanted roofs that were added to shield truss members from precipitation, sparks, or both.***
Pennsylvania State Archives MG-199

In a view taken by Philadelphia photographer William Herman Rau (1855-1920), the eastbound Pennsylvania Limited, *PRR's all-Pullman, extra-fare luxury passenger train between New York and Chicago, poses at the east end of the 1877 iron Rockville Bridge. The train was established in October 1881 as the* New York and Chicago Limited *and the name* Pennsylvania Limited *was adopted systemwide in October 1891.* American Premier Underwriters, Inc.

3. From wood to iron

Ever since 1871, when the Pennsylvania Railroad replaced its crossing of the Juniata River at Mount Union, almost 100 miles west of Harrisburg, with an iron bridge, Rockville Bridge remained the last wooden river span on the main line. A business recession – they were then called panics – hit the nation in 1873, and PRR's freight revenue, after having doubled in the short seven-year period from 1866 to 1873, began to recede. By 1876, the slide was arrested and although freight business did not yet rebound, PRR took advantage of a one-time demand for travel to the nation's Centennial celebration in Philadelphia to reap a boom in passenger business. The economy remained stagnant, which soon led to other problems for PRR, but the company was ready to get back to the job of modernizing its physical property.

The railroad contracted with the Pencoyd Iron Works in Montgomery County (suburban Philadelphia) to supply 700,000 pounds of wrought-iron members with which to replace the old Howe-truss spans.[1] Established in 1852, the firm had begun building iron bridges in 1859, one of the earliest in America to do so. The job of erecting the bridge was given to the Delaware Bridge Co. of New York, a contract that was projected to last just five months, from July 1 to December 1, 1877.[2] Despite overblown newspaper reports that the job would run to $1.5 million, the cost of the work was $326,614.[3]

It was not a simple task. The object was to replace the single-track wooden bridge with a double-track wrought-iron bridge, using the same piers, and doing so while keeping the line open. The first hint of change came in early June, when PRR rearranged the telegraph wires on the wooden bridge to allow construction to begin.[4] Masons soon went to work extending the 1849 piers and widening the two abutments to accept a double-track superstructure.[5] At the same time, work began on widening for a second track the "boiler plate" bridge over the Pennsylvania Canal, just east of the river bridge on the Rockville side.[6]

Wide-ranging labor trouble

On June 1, 1877, PRR announced a 10 percent wage reduction as a result of the continued recession. Other railroads made similar cuts. Work on the bridge was barely under way when a strike that started on July 16 on the Baltimore & Ohio Railroad in Maryland spread across Pennsylvania, leading to mob rioting,

A detail view of Rau's Pennsylvania Limited *photograph shows train and engine crew members posing on and in front of the engine and in the open baggage door of the Pullman combination car* Premier. *The artfully pinstriped engine is No. 568, a D10a-class 4-4-0-type locomotive built in PRR's Altoona shops in June 1891.* Collection of Walt Ready

"Iron, where properly used, is more safe than timber, and I am decidedly of the opinion that none of the bridges on the Pennsylvania Rail Road should be rebuilt with wood – iron should be substituted in every case when renewals are required."

– Herman Haupt, PRR Chief Engineer 1853-1856

River, canal, and railroad intersect at Rockville. Engine No. 64, a Class D3 4-4-0-type locomotive built by the Baldwin Locomotive Works in March 1876, leads an eastbound passenger train. It is pausing for a photo at RJ block station, which exhibits the Pennsylvania Railroad's signature tower design that features an octagon-shaped second-story turret. The light-colored cars behind the engine are mail cars. A canal boat is visible beneath the easterly span of the iron girder bridge.
Railroad Museum of Pennsylvania (PHMC)

and some arson, directed against railroad property in Pittsburgh, Harrisburg, Reading, and other points. Because many railroaders could barely feed their families and keep a roof over their heads, hundreds of train and engine crew members went out on strike, blocking the movement of trains manned by non-strikers.

Pennsylvania Governor John F. Hartranft, traveling in Wyoming when civil unrest erupted, immediately started east, telegraphing orders to his adjutant general, James W. Latta, to call out the National Guard of Pennsylvania to preserve peace and protect railroad property. This was only partially successful, because many soldiers sympathized with the striking railroaders. In Pittsburgh, rioting started on July 19 and continued for five days, resulting in bloodshed and the destruction of hundreds of cars, locomotives, and railroad buildings. Across the state, many rioters and looters were not railroaders but unemployed youths, drifters, and thugs with time on their hands.

Mobs in Harrisburg prevented freight trains from running, but allowed most passenger trains to move, especially if they were carrying U.S. mail. A group of rioters actually captured some 30 or 40 National Guard soldiers who were trying to return to Philadelphia after being turned back at Altoona. Fearing what might happen if they arrived in the mob-controlled city, the soldiers got off a train at Marysville, intending to walk the rest of the way to Philadelphia. Some of them made their way down the West Shore of the Susquehanna, while other crossed the railroad bridge to Rockville. The rioters rounded them up, disarmed them, fed them at a hotel, and put them on a train for Philadelphia.

The worst destruction had taken place in Pittsburgh by July 25, when Major General J.K. Sigfried, commanding the Fourth Division of the National Guard of Pennsylvania, ordered the Eighth Regiment under Colonel J.P.S. Gobin to march from the state arsenal in Harrisburg to Marysville to guard the Rockville and Northern Central bridges. Men were placed on both sides of the river at the ends of both bridges. Only 50 miles east, rioters in Reading had already burned a bridge on the Lebanon Valley line. That same day, Governor Hartranft was aboard an eastbound train en route for Harrisburg, having returned from the West to gain firsthand control of the situation.

The infantry regiment guarding the bridges was made up mostly of men from Lebanon, York, and Schuylkill counties, but also included Company D, the Harrisburg City Grays, commanded by Captain Thomas Maloney. A volunteer 40-member company of militiamen from Wrightsville, York County, joined the others at Rockville on July 28 and re-enlisted there. The tense mood of the previous week had apparently eased, because on Monday the 30th, *The Patriot* newspaper reported: "Quite a number of

Harrisburgers visited the City Grays (Captain Maloney) encamped at the west end of the Rockville Bridge, yesterday." All of the National Guard troops remained on patrol there until the following day, when they returned to the arsenal at Harrisburg to be redeployed to Pittsburgh.[7]

Bridge work proceeds

By August, an observer noted that 40 or 50 men were working on the project, adding: "A large amount of iron is on the ground. Two portable steam engines are to be used for hoisting material. A narrow track runs out on the bridge from the Rockville side of the river, on which material is trucked when wanted. The abutment on the Perry County side is being extended, and some alteration on the Rockville side will also have to be made."[8]

Although they aren't detailed, surviving accounts do reveal something of the staged procedure for replacing one set of spans and adding another, while keeping trains moving during construction: "[The] wooden structure … was torn away as the iron bridge took its place, the trains running for a part of the distance on the wooden bridge and then gliding upon the iron bridge."[9]

By September 2, four iron spans were in place and trains began using them that day, with three more close to being finished. *The Patriot* reported that the Delaware Bridge Co. had 100 men on the job.[10] Work went forward rapidly, because by Saturday, October 20, the same newspaper announced: "Five spans more of the new iron bridge over the Susquehanna at Rockville remain to be built – three on the Rockville and two on the Perry County side of the river. Another span will be completed by Sunday, leaving only four to be finished.[11] Two weeks later, on November 5, *The Patriot* used an underwhelming one-sentence story to report: "The new iron bridge at Rockville was completed yesterday."

Being built on the old piers, the iron bridge was the same length, 3,680 feet, and had the same number of spans, 23, each 160 feet long,[12] as the wooden bridge. Like the wooden bridge, it was a deck bridge, meaning that the track rested atop the superstructure. The design of the superstructure was a lattice truss, somewhat rare. Finishing work went on for some time, including the placement of new telegraph poles and lines. *The Patriot* noted that during construction of the iron bridge, some "six or seven men have tumbled into the river," two of whom died – one was a construction worker and the other was a watchman. Another worker suffered severe injury when his leg was crushed by a heavy car used to move iron onto the bridge.[13]

A rustic view by William H. Rau shows a child posed on a rock outcropping overlooking the west end of the double-track iron Rockville Bridge. In the foreground are the tracks of the Northern Central Railway leading south (right) to West Fairview, Wormleysburg, and Bridgeport (present-day Lemoyne), three communities that lay across the Susquehanna from Harrisburg. At left is the edge of PR Tower.
American Premier Underwriters, Inc.

PRR's Rockville-Dauphin branch

The same causes – longer, heavier, and more frequent freight trains – that prompted PRR to replace the wooden Rockville Bridge also pushed PRR and its Northern Central subsidiary to seek a direct Dauphin-Rockville connection. NCR had established a large freight classification yard at Marysville, along with a roundhouse and machine shop for locomotive maintenance and repairs.

Because PRR gained control of NCR in the 1860s, the two were now working as affiliated companies. With the Pennsylvania using a new double-track iron bridge about two miles downstream, it no longer seemed to make sense to maintain two long bridges over the same river so close to each other. Officials began to consider rerouting trains and abandoning the old wooden Dauphin bridge. It was true that diverting southbound NCR freights to Rockville, over Rockville Bridge and then northward for the brief jog into Marysville Yard would add some running time, but that seemed to be an acceptable tradeoff for the benefits of retiring a fire-prone structure that required constant maintenance. NCR freights had to stop at Marysville Yard anyway for switching and classification before moving south to Baltimore or north

Rockville and Marysville

When trains first started running on the Pennsylvania Railroad in 1849, the community of Rockville, situated five miles upriver from Harrisburg, existed, but Marysville did not. On PRR timetables, the first station listed west of Rockville Bridge was Cove, which was four miles west of Rockville; the next station beyond that was Duncannon, four miles farther.

The Rockville area was first settled about 1725, and during the French and Indian War, the British established Fort Hunter there as a military blockhouse. In the late 18th and early 19th century, Rockville became home to a grist mill and a sawmill, a blacksmith shop, a school, a store, a tavern, and a distillery. With the opening of the Pennsylvania Main Line Canal in 1834, Rockville became a classic canal town, but being so close to Harrisburg, it was not an overnight stopover point for canal travelers. A road along the Susquehanna river front into Harrisburg was established as a toll road, but landowners along the road did not pay tolls. The area took on more importance with the coming of the PRR in 1849 and the Schuylkill & Susquehanna Railroad soon thereafter, and then the PRR Rockville Branch in 1883.

Two postcard views from the stone-bridge era show the community of Rockville. above, ***Front Street leading south from the bridge (the east abutment wing wall is visible at extreme left).*** right, ***The town as viewed from the bridge. The gap between the two houses in front and the row of houses behind is the former location of the abandoned Pennsylvania Canal.*** Both cards, collection of Eric Ready

The area on the west side of the original Rockville Bridge was known by a variety of names, including Haley, Haley's Station, Fenwick, Fishing Creek, and Morleton. The arrival of the Northern Central Railway in 1856 brought more local development than had the Pennsylvania Railroad, because the NCR established a freight classification yard and an 18-stall roundhouse (later extended to a full circle of 36 stalls) for maintaining locomotives. The settlement took on the name Marysville about 1860, and was incorporated as a borough with that name in 1866.

By 1875, Rockville boasted two churches and several hotels, and its population stood at 259. Marysville also contained several hotels and three churches, but it had surpassed Rockville in size, with 863 residents. The reason was that employment opportunities were more numerous, including the NCR enginehouse and railroad yard, a forge, a flour mill, and a door-and-sash factory.

Soon after 1900, trolley service from Harrisburg reached both Rockville and Marysville. As personal transportation switched to autos, both towns found themselves on major highway routes. Marysville was located on the Susquehanna Trail between Williamsport, York, and Baltimore, while Rockville was situated on the William Penn Highway linking Allentown and Harrisburg with Pittsburgh.

Marysville has long adopted a borough form of government, with local elections and representation within Perry County, while Rockville remains an unincorporated village in Susquehanna Township, Dauphin County. Because the boundary between Dauphin and Perry counties is the low-water mark on the west bank of the Susquehanna, all three of the Rockville bridges have been located in Dauphin County, with only the west abutment and west approach lying in Perry County. Today the Fort Hunter Mansion and farm have been converted to a museum that is operated by the Dauphin County Parks and Recreation Department. ★

A postcard view looking west on Valley Street from Lincoln Street in Marysville.
Collection of Walt Ready

Guarding the west approach to the iron Rockville Bridge is PR block station, another classic PRR octagonal two-story wood-framed tower. Collection of Ted Xaras

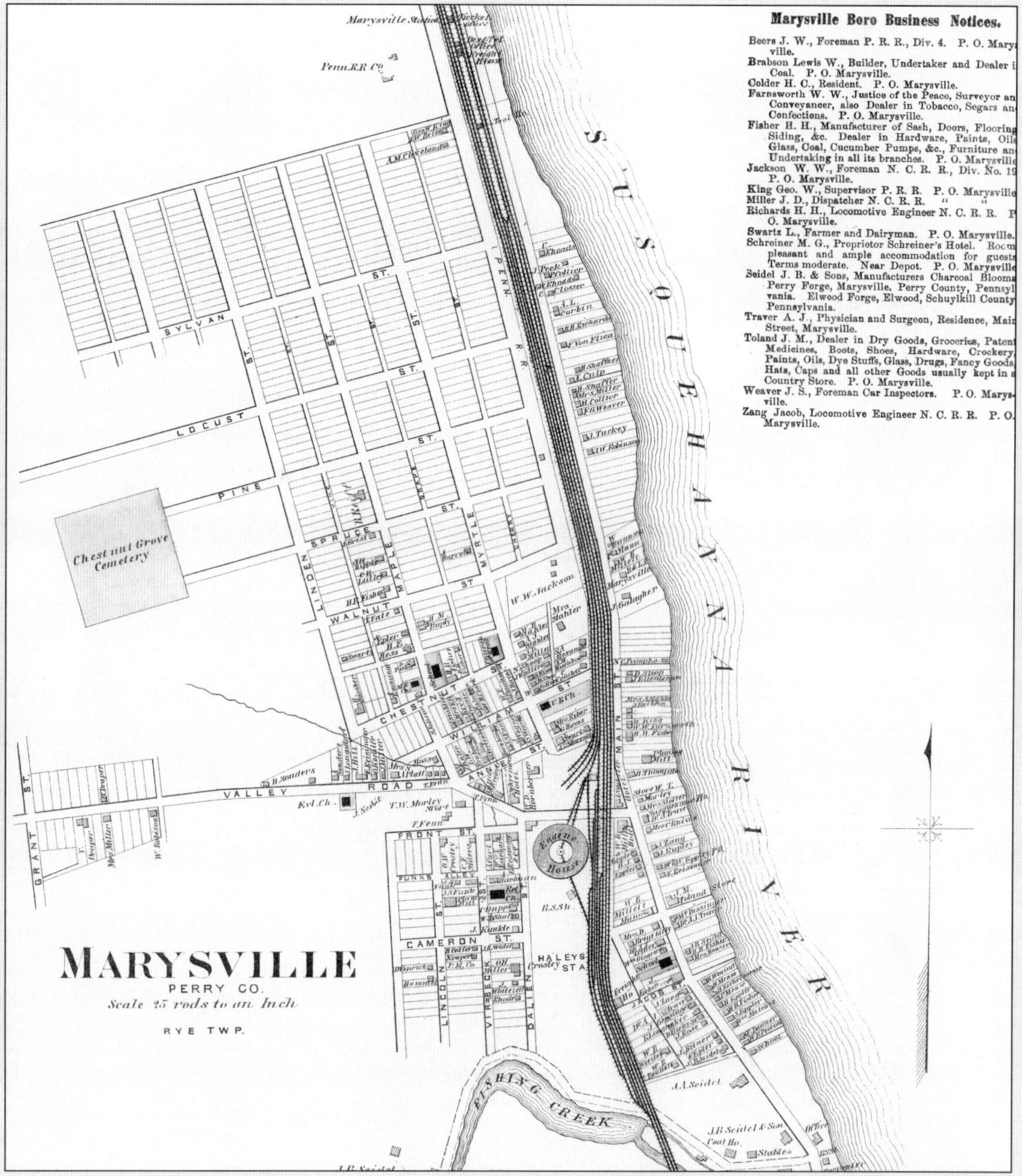

The yards and roundhouse of the Northern Central Railway dominate the center of Marysville in this map from the 1870s.
Collection of Duane Hammaker

to Sunbury, so how they got into that yard became a less important factor. Moreover, by abandoning the old bridge, PRR could remove the level diamond grade crossing in its main line at the north end of Marysville, eliminating a choke point and a potential collision hazard.

With those considerations at hand, PRR in July 1881 reached an agreement with the Schuylkill & Susquehanna Railroad that allowed PRR to lay tracks on part of S&S's right-of-way between Dauphin and Rockville.[14] The first contractor on the 2-1/2-mile branch apparently failed to make much progress, and the contract was canceled and handed over to another, named E.D. Smith, of Philadelphia. In April 1882, Smith had 80 Italian, German, and Dutch laborers at work on the project, along with 50 stone masons and 16 to 20 horses and mules. Blasting away the base of the mountain south of Dauphin was necessary to shoehorn an additional track onto the narrow shelf that was already occupied by the S&S, the canal, and a wagon road. This work was carried on by "James Creswell, of [Harrisburg], with a large gang of colored citizens."[15] Sometimes the dislodged rock went into the canal, which forced the canal workers to draw off water in order to remove the debris. On one occasion, a 600-ton rock landed on the road, leaving just enough room for a wagon to squeeze by.[16] On another, a charge failed to go off and when workers went to drill it out, it fired, killing a black laborer, Charles Collins, and injuring two other people.[17]

An eastbound freight train, headed by what appears to be a Class H3 2-8-0-type locomotive, moves across the 1877 iron Rockville Bridge. Collection of author and publisher

Work on the branch continued through the summer and autumn, and PRR signed an agreement on November 1, 1882, that would hand over the day-to-day operation of the line to Northern Central, its subsidiary.[18] New track connections were needed on both sides of Rockville Bridge: on the east side, to tie the Rockville Branch into the PRR main line, and on the west side, to allow Northern Central freights to directly enter the south end of NCR's yard at Marysville. Final track work was completed on February 5, 1883, and a day later, NCR freights began to bypass the Dauphin bridge, using the Rockville Branch and Rockville Bridge to reach Marysville Yard. The project had cost $175,000, NCR reported to its stockholders, "the funds for which were provided by the Pennsylvania Railroad Company, and a contract made between the two companies, by which your Company [NCR] operates the line and from time to time refunds to the Pennsylvania Railroad Company the outlay thereon."[19]

Abandoned, the Dauphin bridge stood disused for a year. In 1884, the thrifty NCR removed most of it, reusing some of the timbers and sawing up others for lumber. Iron spans that had replaced some damaged or burned wooden spans over the years were salvaged to be re-used in place of wooden bridges on portions of the Northern Central in New York state.[20]

By mid-1883, the Pennsylvania Railroad began to draw up figures on the annual expense of maintaining Rockville Bridge so that NCR could be billed for part of the cost. A preliminary itemized list showed the cost of renewing wooden deck timbers, handrails, and paint, to be $4,333.50.[21] An assistant PRR comptroller wrote to PRR Chief Engineer William Henry Brown, asking for a complete breakdown that included the cost of watchmen. Brown scribbled a note to a subordinate on the margin of the letter before passing it on: "Ask Prevost or [illegible] if there are any watchmen on this bridge."[22] The reply is unknown. Although the superstructure was metal and thus far less vulnerable to fire than the 1849 bridge had been, the deck was still made of wood timbers and was susceptible to locomotive sparks. The erecting of staffed block towers at both ends of the bridge may have rendered bridge watchmen unnecessary, which may have prompted Brown's inquiry. ★

above, ***A panoramic photograph by Frederick Gutekunst shows the 1877 iron bridge from near water level, looking west from the Rockville side. Each span consisted of three lattice trusses made up of repeating X-shaped patterns of bracing, with the vertical strut in the top of each X being an unusual addition. Temporary scaffolds for maintenance have been hung in the west end of the first span. The boat and dock at right are real and are part of the original photograph, but an artist has penciled in the boat at left.*** Railroad Museum of Pennsylvania (PHMC)

left, ***The westbound*** **Pennsylvania Limited** ***pauses for a photo by William Rau at PR Tower on the Marysville side of the bridge. Pulling the train is No. 174, a Class D16a 4-4-0-type locomotive built at PRR's Juniata Shops in Altoona in 1895. The dark green cars with cream trim date this as the 1898-1899 version of the*** **Pennsylvania Limited*****; in 1899, red trim was added to the letterboard above the windows, and in 1902, PRR President Alexander J. Cassatt ordered that the primary color for passenger cars be standardized on tuscan red.*** American Premier Underwriters, Inc.

In a view looking upstream, the individual arch ring stones, or voussoirs, on this arch have been laid and it has been "keyed" with the installation of the topmost row of keystones. With the structural integrity of this arch assured, the wooden falsework center has been removed, affording a clear view of the 1877 iron bridge beyond. Collection of author and publisher

4. From iron to stone: 1899-1900

William H. Brown had been chief engineer of the Pennsylvania Railroad since 1874 and saw firsthand the technological transition from wooden bridges to iron. But as trains got longer, heavier, and faster, even the ability of iron bridges to stand up under the strain came into question. Perhaps nothing illustrated this so dramatically as the December 29, 1876, "Ashtabula Horror," in which an 11-car Lake Shore & Michigan Southern Railway (New York Central) passenger train plunged 75 feet through an iron bridge in Ohio, killing 92 people and injuring 64. The bridge design and the level of maintenance, not the choice of material, were defective, but the incident cast a shadow on iron bridge construction. Two LS&MS civil engineers later committed suicide.

Summarizing public opinion, *Harper's Weekly* magazine editorialized by posing questions for which no answers were at hand:

> "Is there no method of making iron bridges of assured safety? And who is responsible (so far as human responsibility goes) for such an accident – the engineer who designed the bridge, or the contractor, or the builders, or the railroad corporation? Was the bridge, when made, the best of its kind, or the cheapest of its kind? Was the contract for building 'let to the lowest bidder,' or given to the most honest, thorough workmen?"[1]

Brown and other PRR executives were undaunted, believing that iron had proven to be a solid choice if a bridge were well designed and maintained. The iron Rockville Bridge went up just seven months after the Ashtabula wreck. In the 1840s and early 1850s, PRR had built numerous stone bridges on its Western Division, but by the 1870s, that construction method was considered to be obsolete. However, with the growing industrialization of America, demand rose for shipments of coal, ore, iron, steel, and other heavy commodities. More powerful locomotives were needed, and freight trains that once carried 50 tons total were now hauling that much weight in just the first few cars. Brown came to believe that to accommodate speed and weight with permanency on multiple-track lines, a return to stone-arch technology was the best solution.

In 1911, civil engineer Henry G. Tyrrell wrote in his self-published History of Bridge Engineering:

> "The greatest forward movement in the building of stone railroad bridges in America began in 1888, when the Pennsylvania Railroad company under the direction of William Brown, chief engineer, commenced replacing its wood and iron bridges with permanent ones of stone and concrete. Several of these safely resisted the Johnstown flood of 1889 ... for 20 years that railroad has continued rebuilding bridges in masonry, and many of great magnitude have been erected, including those over the Susquehanna at Rockville, Trenton, Brunswick, Coat[e]sville and Shocks Mills, as well as many other smaller ones. The bridge at Trenton has 18 spans, while those at Brunswick, Coat[e]sville and Shocks Mills have 21, 10, and 28 spans respectively."[2]

What appears to be the first stone-arch replacement bridge to be installed on the PRR main line was built in 1877, the same year as the iron Rockville Bridge went up. A three-arch bridge over the Little Juniata River replaced an iron structure that had become too light. The iron spans were re-used for overhead public road crossings elsewhere on the railroad.[3] No more stone-arch bridges were built until 1887, when three were added.

Thanks to the Johnstown Flood, the stone-arch bridge over the Conemaugh River at Johnstown, Pa., was the best-known of that type of structure prior to the coming of the stone Rockville Bridge. Completed in 1888, the seven-arch bridge stood firm the following spring when a dam failure unleashed a 40-foot-high wall of water on the city, resulting in more than 2,200 fatalities. Many people died when they were caught in a mass of burning debris that collected at the unyielding bridge. Five other stone-arch bridges were installed on the main line in the same year as the Conemaugh bridge, including the five-arch, 329-foot-long bridge over Conestoga Creek east of Lancaster, Pa.,[4] followed by six in 1890, four in 1891, five in 1892, three in 1893, and four in 1899.

Brown was the chief advocate for stone bridges. Although they cost more than iron bridges to build, they were more stable, did not need the constant maintenance required by iron joints and girders, and never needed to be painted.[5]

Improving the railroad

When Alexander Johnston Cassatt became president of the company on June 9, 1899, he moved decisively to expand the railroad's capacity even more. Under his direction, alignments were straightened and grades were eased all over the railroad, but especially on the New York-Philadelphia-Pittsburgh main line. Almost this entire route was expanded to four, and in some places, six, tracks.

It was clear that the railroad was straining at the seams. In the summer of 1899, PRR General Manager Joseph Hutchinson wrote to Chief Engineer Brown, asking his opinion on running heavier freight locomotives on the iron Rockville Bridge. "While I know very well your views with regard to running mogul engines over Rockville Bridge," he wrote, "I would be very glad to know, if we limited the number to half a dozen per day, as to whether you think it would have any appreciable effect upon it. Of course it would be a different matter if all of the engines were of this type."[6] In general railroad jargon, the term *mogul* refers to any steam locomotive with a 2-6-0 wheel arrangement of two leading wheels, six driving wheels. and no trailing wheels under the firebox. PRR was operating F1- and F3-class 2-6-0 type engines in fast freight service about this time, but Hutchinson may have had a wider definition in mind. The term *mogul* in PRR circles sometimes referred to 2-8-0 or Consolidation-type engines, which in PRR's classification were H-class locomotives, and occasionally to 4-6-0 engines, which PRR termed the G class. All three types were used in the Harrisburg area. In any event, Brown's reluctance to allow heavier engines on the iron bridge highlights the fact that it had long since reached its design limits.

Economics was driving the trend toward heavier cars and more powerful locomotives. One veteran railroader made the following observations in 1899: The average carrying capacity of a freight car rose from 12 tons in 1870 to 20 tons in 1880 to 40 tons in 1890, and it was edging toward 50 tons in 1899. Thus a train of 40 cars – an average length – in 1870 would carry 480 tons, but in 1899 the same length of train would carry 2,000 tons. On a level grade, a heavy locomotive could pull about 70 of the 40-ton cars, or about 2,800 tons, with a six-member crew. The same amount of freight in 1870 would require 233 cars and seven engines and crews, or about 35 men.[7]

On November 2, 1899, Cassatt and his top officers rolled through Harrisburg aboard a "president's special" on the homeward leg of an inspection trip to the Pittsburgh area, but nothing definite about the company's plans at Rockville, or elsewhere, leaked out during or after the trip.[8]

Rumors of a new bridge

By the end of 1899, some railroad observers around Harrisburg were predicting that PRR would build a new stone-arch bridge at Rockville, while others scoffed at the idea. During the last week of February 1900, contractors began swarming around Rockville to examine the site in preparation for bidding on work,[9] but the nature of that work was still a closely guarded secret. The company's annual report for 1899, released in March 1900, identified the need for change, but disclosed nothing specific: "The bridge over the Susquehanna River at Rockville needs reconstruction ...."[10] This frustrated *The Patriot* newspaper in Harrisburg, which had reported earlier that a new stone bridge would be built. It commented on March 6: "The indefinite language used makes it impossible to tell whether the company intends building ... a new bridge at Rockville or if it is intended but to reconstruct [it]."[11] Two days later, on March 8, the same paper said its railroad sources around Harrisburg were still divided as to what would happen, noting that some observers thought the piers of the iron bridge would be enlarged for a four-track superstructure, just as those of the wooden bridge had been lengthened in 1877 to support the double-track iron spans. It added: "But this is not likely as the cost … would be nearly equal to that of building a stone bridge there over a better route than the old bridge."[12]

The suspense soon vanished. Harrisburg's *Star-Independent* on March 9 reported that a stone-arch bridge of three or four tracks would be built, that it would take several years to construct, and that it would cost "millions of dollars." It identified PRR civil engineer George Nauman (though it misspelled his name as Norman) as the principal in charge, calling the project "one of the most gigantic series of arches yet attempted in civil engineering." Nauman was then overseeing the completion of a second two-track tunnel, 1,075 feet long, through Tussey Mountain on the main line at Spruce Creek west of Huntingdon, Pa., more than 100 miles west of Harrisburg. As soon as his duties there ended, he would transfer to Harrisburg.[13] The report erred in one respect, saying that the bridge would contain only 16 arches, not the 48 that were ultimately built.[14] In fact, because the initial contracts covered only eight arches on each side of the river, the paper's count was correct for what it covered, but it was incomplete.

Just a day later, PRR came out in the open, advertising for bids. The first 16 stone arches were to be finished by December 31, 1900, with the successful contractor or contractors having an option to build more arches if they completed the initial ones satisfactorily and if the weather during the winter of 1901 proved to be favorable. PRR purchasing agents began to buy land, not just for the new bridge approaches, but also to enlarge the PRR yards stretching from Harrisburg to Rockville.

As soon as the plan became public, residents of Rockville began to wonder aloud whether the lower, narrower openings afforded by stone arches would cause ice gorges to back up during spring thaws, threatening property on both sides of the river. Rumors were floated that residents would seek an injunction from the state attorney general to block construction, but they never materialized.[15] Despite the talk of legal action, preparations went forward. On the west side, material began to arrive, and workers' sheds were erected near Marysville.

THE PATRIOT had the last laugh

War of the newspapers

When the stone Rockville Bridge was being built, Harrisburg citizens enjoyed a choice of three daily newspapers: *The Patriot*, a morning paper whose editorials were aligned with Democratic politics, *The Telegraph*, an evening paper that took the Republican point of view, and the *Star-Independent*, another evening paper whose name accurately and proudly described its lack of political affiliation.

Harrisburg had long since become a hub for railroad activity, with tracks leading out of the city in eight directions, and railroad employment ran into the many thousands. As a result, all three papers covered railroad activity intensively, including such items as crew assignments, accidents, inspections by railroad officials, the arrival of any new or unusual locomotives or trains, and the passage through Harrisburg of politicians or celebrities aboard this train or that. On some days, the amount of railroad news took up most of a page, sometimes exceeding the amount of sports news.

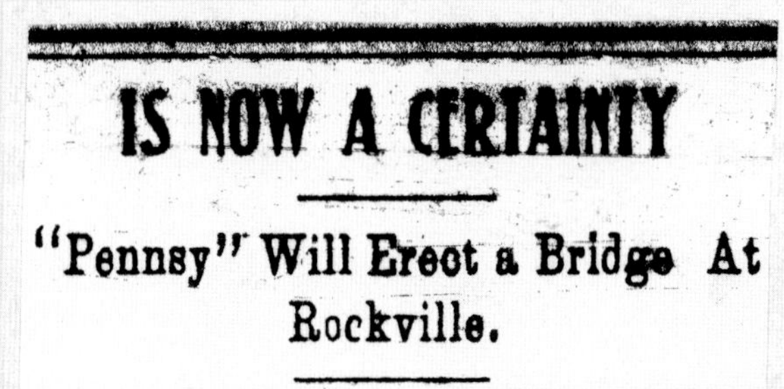

IS NOW A CERTAINTY

"Pennsy" Will Erect a Bridge At Rockville.

The Patriot, *October 30, 1899*

Each paper viewed this as prime competitive territory and tried to "scoop" the others on important developments, including the construction of a stone Rockville Bridge.

On Monday, October 30, 1899, *The Patriot* proudly announced: "The Patriot now has it from excellent source that the Pennsylvania Railroad intends reconstructing its long bridge at Rockville across the Susquehanna River. Some time ago it was printed that there was a rumor afloat to the effect that the structure was to give way to one of more modern design and that rumor has developed into a fact, and it is now known definitely that the bridge will be rebuilt. As stated, the new bridge will be of stone instead of iron … and will be much broader in order to accommodate the increased traffic of the Middle Division."

The same afternoon, the *Telegraph* shot back: "Middle Division officials laugh at the story published by a morning contemporary about the reconstruction of the Rockville Bridge. They say that the present bridge is good for years to come."

Three days later, the *Star-Independent* chortled: "The fake tale about the rebuilding of the Rockville Bridge is being copied by a number of papers throughout the state. Corrections will be in order shortly." Two days later, it took another shot: "Read the Star-Independent for railroad news, not fake stories."

The Patriot held its tongue, but a few months later, on February 12, 1900, added this zinger to an otherwise routine railroad story about a new coal wharf: "Some of the papers that poked fun at The Patriot's story of a four-track bridge at Rockville are now trying to catch up by saying that the numerous improvements in the local yards will necessitate that move within a short time." ★

Contracts are awarded

On March 13, just three days after announcing the project, PRR Chief Engineer Brown awarded contracts for the first two segments of bridge construction. The east side work was won by Drake & Stratton of Philadelphia and Pittsburgh, and the west side work was won by Harry S. Kerbaugh & Co. of Philadelphia.[16] Both had performed many jobs for PRR: Drake & Stratton had assembled the Delaware River Bridge at Frankford Junction on the Philadelphia-New York main line in 1896, and rehabilitated the piers of the Columbia Bridge about 25 miles downriver from Harrisburg to prepare them for a new superstructure in 1897. The extremely short time between advertising and award suggests that PRR officials already knew to whom they would give the work. PRR also reportedly offered a cash bonus to the contractor that finished its work first (a newspaper later refuted this), and the promise of awards of additional contracts on the project. Construction was set to begin on April 1, and the final cost was expected to be $1 million.

"The Rockville Bridge, when once constructed after the plans already adopted, will be a model of strength and beauty, and it is estimated that it will last for a century with ordinary repairs."
– Harrisburg Star-Independent, March 15, 1900

The bridge would be 3,791 feet long between abutment walls, and about 3,860 feet long overall, with each arch spanning a distance of 70 feet, carrying four tracks 46 feet above low water level. At the height of the coping (outside top walls), the width of the bridge would be 52 feet. Its alignment was level and, for the most part, straight. At the last four arches on each end, the track curved in a sweeping arc to meet the alignment of the mainline approaches.[17]

The stone bridge crossed the river almost perpendicularly to the current, while the iron bridge and its predecessor had crossed at a 68-degree angle. On the east side, the new bridge was located about 40 feet south of the iron bridge, and on the west, it made landfall 500 feet south of the old bridge. Many armchair observers wondered why the new bridge crossed at a diverging angle, rather than running parallel to the old bridge. The reason appeared to be twofold. First was PRR's intent to build a connecting track on the west side that would extend southward to the Northern Central Railway main line. This would allow NCR freight trains from Baltimore and York to proceed northward to Sunbury, Williamsport, Erie, and Buffalo without going into Marysville Yard and reversing direction. Second, it was easier, faster, and cheaper to chisel and fit the pieces for a straight bridge than for one on which the piers stood at a skewed angle, especially one this long. Either way, the openings beneath the arches needed to be close to perpendicular to the river channel to reduce friction from the passage of millions of gallons of water each minute. Because only the piers, not the superstructure, of the 1849 and 1877 bridges were made of stone, that hadn't been a consideration when those bridges were built.

In planning for the track connections, PRR allowed more width at both ends of the bridge than was needed for the four mainline tracks. On the east, the railroad planned to retain its connection to the Rockville Branch to permit Northern Central trains to and from the north to cross at Rockville in lieu of the abandoned-in-1883 NCR bridge at Dauphin. On the west, PRR made provisions for a new connection with the Northern Central line through Fairview (Enola), West Fairview, and Bridgeport (Lemoyne) to New Cumberland, York, and Baltimore. In retrospect, it's also apparent that this was part of a PRR strategy that included building an immense freight classification yard at Enola, but this was 1900, and the railroad didn't disclose those plans until mid-1902, after the bridge was opened.

Six units of eight arches

Designed as six sections of eight segmental arches each, the bridge employed a thicker or abutment pier at the junction of each two sections of eight. The design is called segmental because the curve of each arch does not blend smoothly into the vertical walls of the piers supporting it, but instead makes a distinct joint at the top of each pier. Forty-two of what PRR termed "ordinary" piers were 10 feet thick at the base (narrowing to eight feet at the "springing line," or bottom of each arch), but the five abutment piers were 21 feet thick at the base (narrowing to 19

General arrangement of arches and abutment at west end of bridge.
Engineering News, May 10, 1900

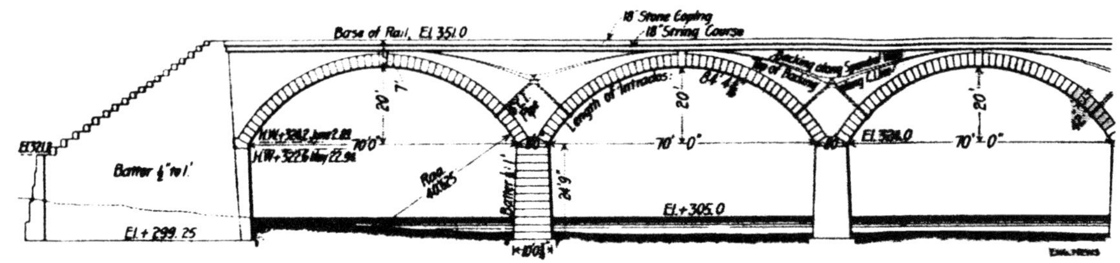

FIG. 1.—PART SIDE ELEVATION OF 3,820-FT. STONE ARCH BRIDGE FOR THE PENNSYLVANIA R. R. AT ROCKVILLE, PA.
Wm. H. Brown, Chief Engineer, Pennsylvania R. R.
Drake & Stratton and H. S. Kerbaugh, Philadelphia, Pa., Contractors.

Life on the railroad, 1900-1902

The transition between the old and new bridges was so smooth that the rhythm of life on the railroad hardly even skipped a beat.

That pattern went something like this: Coal trains moved all year long, coming from mines in the north and the west to supply industries with metallurgical coal for steelmaking and cities with home heating fuel. Preference freights – those carrying merchandise and perishables, including fruit, vegetables, and meat in refrigerator cars – also moved all year long. When the volume of freight overwhelmed the Harrisburg yards' ability to handle it, PRR dealt with it in two ways. First, it temporarily stored coal trains on sidings along the main line from Marysville west as far as Juniata County, about 30 miles west of Harrisburg, while continuing to move the higher-class freight. When the congestion eased, crews were sent out to bring in the stored trains. The second method of working off an overload of business was to add more Sunday crews in the Harrisburg yards until the glut was gone. The company also handled seasonal freight. Each fall, the fruit harvest began to move by rail, including apples and peaches.

On the passenger side of the business, PRR ran its regular schedule of daily east-west through trains connecting the East Coast with Harrisburg, Pittsburgh, and the Midwest. Over Rockville Bridge, this amounted to 19 through trains and six locals each day. These crossed paths with a smaller number of north-south trains connecting Buffalo, Erie, and Williamsport with Harrisburg, York, Baltimore, and Washington. Local all-stops commuter trains that crossed or passed Rockville Bridge included a run between Harrisburg and Altoona, and the Lykens Accomodation, which brought people into the city from points in northern Dauphin County each morning and returned each evening. When the Legislature was in session, which was every other year at that time, senators and representatives rode trains to and from Harrisburg. Trains were also full at the beginning of each session of Dauphin County Court, and for such extraordinary occasions as the executions of condemned criminals.

During summer, area residents used trains to find relief from the heat in cooler surroundings. Several times each season, PRR offered beachgoers special fares on excursion trains to Atlantic City, N.J., as did its competition, the Reading. In addition, Perry County along the main line west of Harrisburg had become a popular place for building summer cottages. Some families moved to these temporary residences for the season while the breadwinners continued their jobs, taking the train into Harrisburg each morning and back again each evening from Marysville, Perdix, Cove, Cove Forge, Duncannon, Juniata Bridge, Aqueduct, Losh's Run, Bailey, and Newport.

For PRR employees, the pay car came around every month and workers lined up to be paid for their time, sometimes by check, sometimes in cash. Evidence of the Progressive movement to improve the lives and livelihoods of everyday workers also began to show up. After several delays, by 1900, federal law required railroads to equip all trains with automatic knuckle couplers and air brakes, safety measures that many in the industry had fought because of the cost. The benefit was that brakemen no longer needed to fear amputation or death from stepping between cars to perform coupling or uncoupling. Another sign of the Progressive drive was that, the same year that PRR started construction of the stone Rockville Bridge, it instituted the Voluntary Relief Fund, an early form of jointly funded health, retirement, and death-benefit insurance. Along with this, it imposed mandatory retirement at age 70 for all employees, up to and including the railroad president, whether or not they chose to join the fund. ★

Harrisburg railroading statistics for 1902

Number of freight cars handled by PRR through Harrisburg: 1.7 million

Number of daily PRR passenger trains: 99

Number of PRR shop employees: 2,126

Total number of PRR employees: 4,707

Number of Reading employees: 2,000

Source: The Patriot, *August 3, 1912*

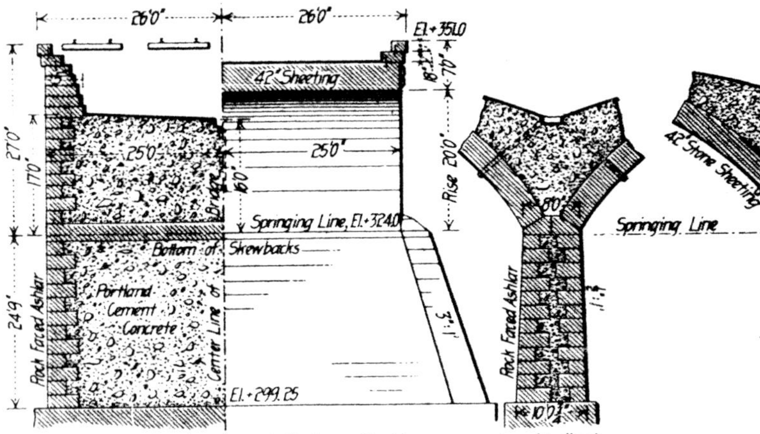

FIG. 4.—SECTIONS SHOWING REGULAR PIER MASONRY, ROCKVILLE BRIDGE.

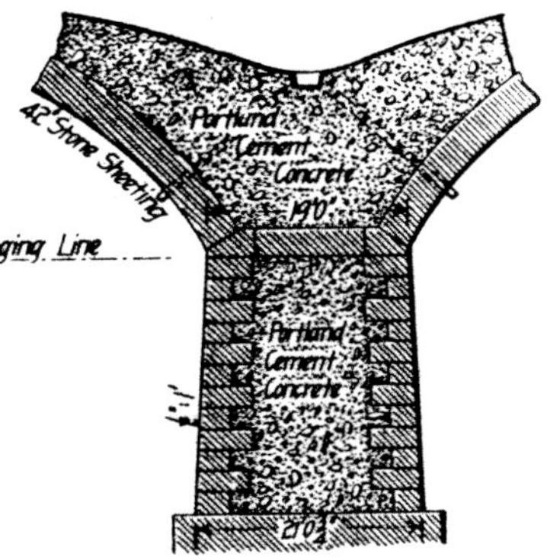

Fig. 5.—Transverse Section Showing Abutment Pier Masonry, Rockville Bridge.

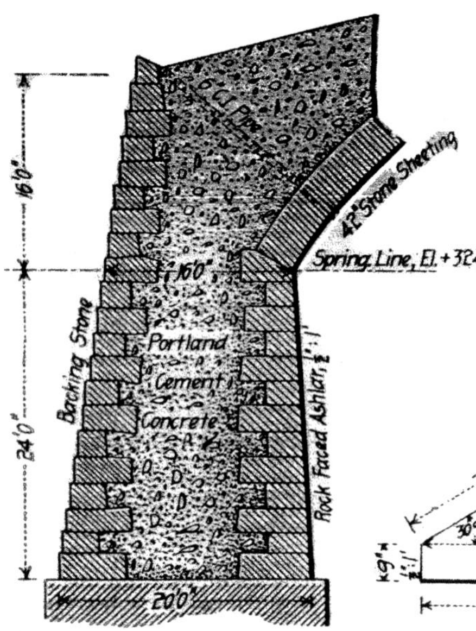

Fig. 3.—Transverse Section Showing Abutment Masonry, Rockville Bridge.

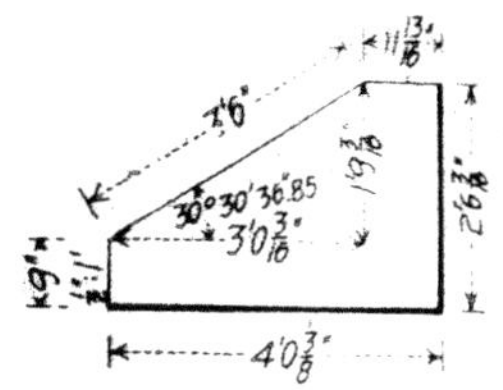

Fig. 2.—Details of Skewback Stones for Rockville Bridge.

above, ***Diagrams showing cross-sections of the regular piers (of which 42 were built) and the double-width abutment piers (of which five were built) show that the centers of all are filled with concrete, making the bridge a composite and not strictly a stone-arch structure.*** left, ***The walls of the abutments at the east and west ends were constructed to this pattern. In both diagrams, the term "42-inch stone sheeting" refers to the 46 rows of voussoirs, or stone segments, that make up each arch. Skewbacks are the triangular-sectioned stones at the top of each pier from which the arches rise.*** Engineering News, May 10, 1900

feet at the arch level). Because each pier supports the dead weight of the two arches that meet at the pier, as well as the moving load of trains above, the use and positioning of abutment piers had two purposes. First, it facilitated the building of such a long bridge by dividing the work – and limiting the potential load from the arches under construction – into segments. This allowed convenient stopping places, as the engineers knew that construction would extend over two seasons. Second, once in service, this design limited the potential damage from any domino-like effect if a catastrophic failure occurred in one or more arches, such as during a flood.[18] Each pier was built with a cutwater, or angled upstream extension jutting into the flow of water. The purpose was to smoothly split the current and, in winter, to deflect and minimize damage from moving ice.

Like many other stone-arch bridges along the PRR main line, this one was built of white or pinkish-white sandstone that came from a variety of quarries, mainly in Cambria and Clearfield counties in western Pennsylvania. The stone was cut in 24-inch-thick blocks for the piers, and in 18-, 20-, and 22-inch-thick blocks for the spandrel walls (the "sides" of the bridge that run parallel to the track, that is, the walls above each pier and between the arches). The resulting square-hewn, cut and dressed blocks of stone are known as ashlar. Above track level, forming an outside wall on each edge and projecting out over the face of the piers and spandrel walls, are two more rows of stone – one called a string course, and the highest row, called the coping. Rising 20 feet across its 70-foot span, each arch consists of rows of 46 voussoirs, or arch-ring segment stones, and one row of keystones, which are the topmost voussoirs in each arch. A total of 100,000 cubic yards of masonry was required.[19]

The bridge is a stone-arch design, but only on the exterior. Hollow sections in the middle of each pier were filled with Portland cement concrete, as were the areas between the spandrel walls. An initial contract was awarded to the American Cement Co. to provide 150,000 barrels of cement, about a fourth of the expected total.[20] Cement was also spread between the blocks to serve as mortar. Two iron drainage pipes were installed above each pier.

A PRR civil engineer named H.S. Righter prepared the overall plans, but working diagrams were drawn up by a civil engineering corps whose members worked in a field office converted from a former candy store that stood along the Fort Hunter Road (Front Street) at Rockville. The size of this crew is unclear, but very early in the project, it was enlarged by eight men.[21]

Quarries Supply Sandstone

Contractors Drake & Stratton and H.S. Kerbaugh were both thoroughly familiar with using western Pennsylvania sandstone in stone-arch bridges, abutments, underpasses, tunnel portals and linings, retaining walls, and even buildings. Most of the sandstone used in the Rockville Bridge came from Cambria and Clearfield counties.

After getting the go-ahead to start work, the two contractors opened a quarry on Laurel Hill above Conemaugh Furnace, near Johnstown, Pa., to supply stone for the project, as well as for PRR's new Pittsburgh passenger station (Kerbaugh) and the renovation of the original 1850 double-track Spruce Creek Tunnel (Drake & Stratton). A steep 1,475-foot-long inclined plane was built to lower stone from the quarry pit, where 200 men were employed, to the level of a PRR siding for shipment.[1]

Drake & Stratton had already acquired ownership of Bloom's Run Quarry, northwest of Curwensville, Pa.[2] That Clearfield County town boasted two other sandstone quarries and a sizable population of stone masons that had emigrated from Italy. At their peak, the three quarries employed 600 to 900 men and produced a durable, all-weather building material that was used in the Philadelphia Museum of Art, the University of Pittsburgh's Cathedral of Learning, and Harrisburg's Market Street Bridge.

All three quarries blasted, cut, and finished stone for the Rockville Bridge project over its two-year duration. Some Curwensville stonemasons traveled to the bridge site to work there. According to a history of that industry, a mason named Dominic Stefano cut the keystones for each of the 48 arches at Bloom's Run and personally keyed the arches (set the keystones in place).[3]

Even these four quarries were inadequate to supply the estimated 100,000 cubic yards of masonry needed for the bridge, which far exceeded by several orders of magnitude – the size and scope of any previous PRR stone-arch bridge. Other, later, reports note that stone was also obtained from Beaver, Indiana, and Westmoreland counties, which are also located in western Pennsylvania.[4]

In all, the stone used for the bridge amounted to 220,000 tons. ★

Gondolas of cut sandstone have arrived on the Marysville side of the bridge and are being unloaded by steam-powered wooden booms. Collection of author and publisher

On March 20, 1900, a PRR special train traveled from Philadelphia to Pittsburgh, carrying PRR President Cassatt, three vice presidents (including future PRR President Samuel Rea), Chief Engineer Brown and other officials. They were planning to inspect PRR lines from Greensburg to Uniontown, Pa., and a site in Pittsburgh, where a new passenger station was being built, but the train slowed as it passed the construction area at Rockville, already abuzz with workers and piles of material on the ground. Officials crowded the open rear platform of their observation car to get a better view.[22]

They were not the only passengers who were eager to view the new construction. *The Patriot* reported: "There is not a passenger on any of the trains of the Pennsylvania passing that point too tired or sleepy to wake up and take a passing glimpse at the great work just begun."[23]

Two horses are serving as propulsion to move a standard-gauge gondola loaded with timber at the west-side staging area. Lumber in all forms will soon be needed for many purposes, including the placement of cofferdams in the river and a narrow-gauge pile-trestle construction railroad. Except as noted, all construction photos from collection of author and publisher

Almost lost in the excitement over the stone bridge was the fact that PRR also hired a contractor, Sparks & Evans of Philadelphia, to strengthen the iron bridge for the duration of construction, which was variously estimated to take two to four years.[24]

With employment rising, PRR added coaches to a twice-daily "relief," or workers', train that customarily ferried area PRR employees around the Harrisburg area to and from their jobs. It was expanded from three cars to five[25] and extended its trip to Marysville.[26] Timber and lumber began to arrive at both ends of the construction site. Timber was used for building narrow-gauge railroad trestles from each shore out into the river, about 15 feet above the water lever, by which stone would be carried to each pier location. Lumber was used for building tool houses, work platforms, and workers' shanties, and for cofferdams to hold back water at the excavation site for each pier foundation.

The work begins

As the expected April 1 starting date approached, the contractors began to hire every available man, including Northern Central Railway employees from the Marysville yards, former boatmen from New Buffalo and Liverpool along the canal, and farmhands. Local farmers worried that they might have to pay high wages to attract summer help.[27]

Surveys and preliminary grading got under way. Sidings to receive carloads of material were connected to the main line at Rockville and Marysville. Civil engineers were out in boats taking depth soundings as preparation for building the pier foundations.[28] Contractor Drake & Stratton transferred some employees from the Spruce Creek Tunnel job, a fact that worried local residents and police, because the work camp there had earned a nasty reputation as lawless territory. Robberies, holdups, shootings, and stabbings were common, and one account said 18 workmen had been murdered in the course of the work.[29] At least 75 men from Spruce Creek were transferred to the bridge work at first, and more were expected when the tunnel was opened (which occurred on May 6).

Accustomed to quiet rural life, area residents saw both drawbacks and benefits coming their way: "It is said that nearly every house in Rockville and Marysville will be occupied by working-

William Henry Brown, PRR Chief Engineer 1874-1906

"The Supreme Court of the Pennsylvania Railroad"

As chief engineer of the Pennsylvania Railroad for nearly 32 years, William Henry Brown presided over the complete rebuilding of the company's New York-Pittsburgh main line, including construction of the stone Rockville Bridge.

A native of Lancaster County, he was born on Leap Year Day, February 29, 1836. Known as "Billy" Brown in his youth, he was the grandson of Congressman Jeremiah Brown, who represented Lancaster County from 1840 to 1844. He was educated at Central High School in Philadelphia, after which he took up surveying as a prelude to civil engineering. In October 1861, at age 25, he ran preliminary survey lines for the Lancaster, Oxford & Southern Railroad, a projected narrow-gauge line, in Lancaster County. During the Civil War, he served with Colonel Thomas A. Scott's military organization to build and repair Union railroad and telegraph lines.[1]

William Henry Brown
From William Bender Wilson's 1895 PRR history

When Scott returned to civilian life, so did Brown, and he served in a variety of positions of increasing importance with PRR and its subsidiaries through the 1860s, including construction of maintenance of way and car repair shops at the railroad's principal shops city of Altoona, Pa. In the 1870s, he was resident or chief engineer for PRR divisions or subsidiaries until he became superintendent of the Lewistown Division and then the Bedford Division. From there, he was appointed to the position of PRR chief engineer on August 1, 1874, although because of an odd organizational arrangement, he was not given that title until 1881.

He worked to rebuild the railroad following the labor riots of June 1877 and the Johnstown Flood of 1889, and oversaw the construction of Broad Street Station and general office building in Philadelphia, Union Station and general office building in Pittsburgh, and Harrisburg station. Other projects included the Frankford Bridge over the Delaware River, grade crossing elimination in New Jersey and Philadelphia, and the replacement of dozens of iron bridges along the main line with the signature stone arch bridges that became so closely identified with the Pennsylvania Railroad. He oversaw the rebuilding, realigning, and four-tracking of the main line that took place during the administration of PRR President Alexander J. Cassatt.[2]

Despite having such sweeping responsibilities, he was capable of hands-on action. When PRR in 1902 proposed to replace grade crossings in Middletown, Pa., with subways (underpasses) and build a new passenger station, Brown and project engineer George Nauman personally attended a borough council meeting to explain the plans.

Brown earned a reputation for aggressiveness. When he stepped down in 1906 after reaching PRR's mandatory retirement age of 70, the *Lancaster Inquirer* newspaper wrote: "Mr. Brown has frequently shown his impatience of delays created through court proceedings, and it has not been unusual to find that, anticipating such delays, he had gone ahead quickly and accomplished what he was after before he could be interfered with. It has happened that eminent counsel of the Pennsylvania Railroad came to him to discuss the propriety of certain contemplated action or to advise him such and such a thing could not be done, only to be informed by him: 'But I have done it.' The result was that he became known as 'the Supreme Court of the Pennsylvania Railroad.'"[3]

He died in Belfast, Ireland, on June 25, 1910, at age 74. ★

George Nauman, Civil Engineer

Part of a Pennsylvania Railroad Family

The civil engineer in charge of overseeing construction of the 1902 Rockville Bridge was George Nauman, a native of Lancaster County. Other Pennsylvania Railroad projects he supervised were the addition of a second Spruce Creek Tunnel on the main line west of Huntingdon, Pa., and the renovation of the original 1850s tunnel at the same place; the construction of Enola Yard near Harrisburg (1903-05); the remodeling of Harrisburg station (1902); the elimination of grade crossings in Harrisburg and Middletown, Pa. (1902-03), and the construction of the low-grade freight line from Enola to Columbia and Parkesburg, Pa. (1905), which included another massive stone-arch bridge over the Susquehanna, the 2,209-foot-long, 28-arch double-track Shocks Mill Bridge.

George Nauman Collection of Ellen Martin

One of eight children, Nauman grew up in Lancaster as the son of another George Nauman, a lawyer who had once run for Congress against the abolitionist statesman Thaddeus Stevens. Family tradition holds that the younger Nauman was so bright that he entered Lehigh University at the age of 16. He graduated from Lehigh in the Class of 1890 and spent much, if not all, of his career with PRR. Born with congenital cataracts, he eventually lost his sight. He was a lifelong bachelor. His last post with PRR was in the Pittsburgh area in the 1920s, and he died in his 50s.

George's brother Spencer G. Nauman moved to Harrisburg about 1915 to practice law with the widely known attorney Lyman Gilbert. Their firm represented PRR locally for many years, and Spencer G. Nauman's son Spencer G. Nauman Jr. also entered law and joined the firm. Among the cases that the firm handled for PRR were issues with the Public Utility Commission, the Full Crew Law of the 1930s and again in the 1950s, and settlement for claims arising from death and injuries suffered in the July 1962 derailment of the Phillies Baseball Special at nearby Steelton, Pa. Now named Nauman, Smith, Shissler & Hall, the firm still represents Norfolk Southern Corp., owner of most former PRR lines, on local railroad issues.

George and Spencer G. Nauman Sr.'s sister Elizabeth married Porter Allen, who became a PRR regional vice president. Spencer G. Nauman Jr. commented on the railroad connections within the family: "My father was always envious of [his sister]. She could ride the *Broadway* [*Limited* on a pass] and he couldn't." ★

men at the bridge for board and lodging," reported the *Star-Independent*. "A gang of colored workingmen secured boarding at the house of a colored resident of Rockville, located up in the mountains some distance from the town. Both joy and [dis]pleasure have been expressed by the residents of Marysville and Rockville at the project that is bringing so many men to these towns. It is anticipated that a great amount of disorder will follow. It is anticipated that a great amount of money will be left in both places by the workingmen on the bridge."

By April 4, one paper reported, nearly 100 foreign workers were on the job, and "nearly every train that stops at Rockville unloads its quota of laborers for the Rockville Bridge."[30] PRR bought many properties on both sides of the river, both for right-of-way, and for housing for supervisors and track workers.[31] Besides blacks, Austrians, Hungarians, and Italians were soon working on the project, with one observer writing, "The Austrians are competent derrick handlers."[32] The *Telegraph* reported: "Almost every day a new gang of Italian laborers arrive and is put to work."[33] Some Italians were laborers, but others were stone masons who skillfully used chisels to perform final cutting and fitting.

Blueprints for the masonry were being drawn up as dirt was already flying.[34] Less than two weeks into the job, excavation for the west abutment, on the Marysville side, was under way.[35] Kerbaugh's job was much more involved than Drake & Stratton's because it involved slicing away part of the nose of Blue Moun-

Men and boys pose at the fabrication site of one of the wooden centers used to lay the stone arches. Collection of Mel Baker

The Contractors:
H.S. Kerbaugh, Inc. and Drake & Stratton

Although Harry S. Kerbaugh & Co. and Drake & Stratton built many large projects for the Pennsylvania Railroad and other clients, little evidence remains of their business activity.

Both had offices in Philadelphia. About the same time as it was working on Rockville Bridge, Kerbaugh established a plant to maintain its equipment at Bellwood, Pa., east of Altoona, which employed 500 workers. In addition to doing railroad work, the firm built municipal and state improvements and state highways and bridges. Its best-known project was the $14 million Kensico Dam and Reservoir near White Plains, N.Y., completed in 1915, which is a main source of water supply for New York City. It impounds water from sources in the Catskill Mountains and the Delaware River. In 1910, Kerbaugh also constructed a part of the New York State Barge Canal for $2.9 million.

Kerbaugh, also known as H.S. Kerbaugh Inc., became a subsidiary of a New York firm named Kerbaugh-Empire Co. and existed as late as 1926, when the latter firm was a defendant in an Internal Revenue Service suit over deductions to income for losses in 1913, 1914, 1916, 1917, and 1918 that landed in the U.S. Supreme Court.[1]

In addition to a Philadelphia office, Drake & Stratton maintained one in Pittsburgh and one in Duluth, Minn., but most of the company's activity seems to have taken place in the East. In the late 1880s, a partnership of John H. Drake and William D. Stratton was buying properties near McKeesport. That the firm did much business around Pittsburgh also is suggested by an accident account from 1900. The company was building a pier for a Union Railroad bridge over the Monongahela River near Rankin, Pa., when one of its portable boilers blew up, injuring 13 men.[2] Drake & Stratton also was contractor on the Baltimore & Ohio Railroad tunnel beneath 25th Street in Philadelphia.

Both firms owned locomotives for construction purposes. In addition to narrow-gauge steam "dinkies" such as those used in building Rockville Bridge, both owned other kinds of engines. Kerbaugh owned at least 14 Shay geared logging locomotives in partnership with another firm that operated in Oregon and Washington state, while Drake & Stratton in 1888 bought new 2-6-0-type locomotives. ★

Myths or facts:

Longest stone-arch bridge?

From the day it was opened, the 1902 Rockville Bridge has been touted as either the longest stone-arch bridge in the world, or the longest stone-arch railroad bridge in the world. Neither claim is 100 percent right, because strictly speaking, it isn't just a stone-arch bridge. Because its piers and the area between spandrel walls are filled with concrete, it is a stone-and-concrete arch bridge.

In fact, Henry Grattan Tyrrell's landmark 1911 History of Bridge Engineering covers it twice, once in a chapter on "modern stone bridges" and once in a chapter on "solid concrete bridges." According to bridge experts, it does seem safe to say that it's the longest stone-and-concrete arch bridge in the world, but that title isn't as easy to say or remember as the shorter, but less accurate, version.

Patriotic arches?

Central Pennsylvania folklore has long included the claim that Rockville Bridge contains 48 arches in honor of the 48 contiguous United States. Civil engineers might look at the specifications and comment that the 70-foot span of the arches was about the same as those on all the other stone-arch bridges built by the Pennsylvania Railroad at that time – in other words, the company didn't alter specifications to artificially produce a count of 48. But perhaps the best point to refute this myth is that in 1900, the United States consisted of only 45 states. From 1896 to 1908, the U.S. flag boasted only 45 stars, as Oklahoma, Arizona, and New Mexico were still territories. The latter two did not join the Union until 1912.

tain to make way for a wider sweep of the mainline curve at the west end of the bridge. Within the first month, Kerbaugh had about a dozen steam derricks set up to handle the excavation.[36] Even with a larger task at hand, reporters consistently noted throughout the project that Kerbaugh was further ahead on its work than Drake & Stratton.

As employment rose, the fears of disorder seemed to subside. *The Patriot* reported: "The men who are at work on the bridge do not seem to be quite so wild as they were made out to be by reports from Spruce Creek. There has been some disturbance, to be sure, but not as much as characterized the work on the new tunnel."[37] One reason seems to be that, contrary to the job at the remote Spruce Creek, where conditions were so chaotic that even sheriffs were warned to stay away, the contractors policed their employees more tightly. When a laborer drew a knife on a rigger over a crap game, the rigger knocked him down with an iron rod. The contractor paid off and fired the laborer.[38]

On the east side, one-story boarding houses for workers were built near the abutment. Each could house 16 men, and by May, at least seven had been put up. Tenants paid the contractor $1 per month rent, which included use of a stove and free coal. At that time, laborers made about $1 a day. A new two-story house for the Drake & Stratton project superintendent and his assistants was also erected, containing offices, a dining room, and seven or eight sleeping rooms. It was nicknamed "Hotel Norman" (Nauman) in honor of PRR civil engineer George Nauman.[39]

The first stone was laid on May 1, 1900.[40] During the second week of May, 20 carloads of stone arrived, and some piers had risen to the arch level, while excavation was under way on others. Flat boats with steam hoists were in the river, and pile drivers were installing the wooden pile trestles on which narrow-gauge construction railroads would ferry stone out to the pier sites.[41]

800 men at work

Low water in the river, which is shallow at normal stage anyway, made working conditions ideal. Blasting was necessary on the west side to cut down the bedrock so that stonework

text continues on page 44

A Kerbaugh foreman stands on the wooden trestle that supported narrow-gauge tracks that carried stone for building the piers.

ROCKVILLE VIADUCT

MANY WORKINGMEN ARRIVING.

Valuable Property Purchased to Make Way for the Terminus.

MANY NEW TRACKS BEING LAID.

Nearly every train that stops at Rockville unloads its quota of laborers for the Rockville bridge. There are already about one hundred foreign workingmen on the ground assisting at grading for the side tracks that are being run up to a point near the Rockville landing. Tool houses have already been built here and many other evidences that a great railroad contract is about to be carried out are apparent, both on the Marysville and Rockville ends of the proposed new bridge.

The Rockville terminus of the bridge will be located about forty feet below the present steel bridge. The Pennsylvania railroad company has bought the property of Mrs. J. W. Free, which is located below the present bridge. The

Harrisburg Star-Independent, ***April 4, 1900.***

MEN HARD AT WORK

Contractors on West Side of River in the Lead on Bridge.

ROCKVILLE NOW A MECCA

Above everything else talked of in this city at present is the new bridge across the river at Rockville. The immensity of the task of the contractors is appreciated and daily there are scores of visitors from this city on the scene observing the preparations under way. The report of the big viaduct has spread all over the country and there is not a passenger on any of the trains of the Pennsylvania passing that point too tired or sleepy to wake up and take a passing glimpse at the great work just begun.

On the west side the contractors are apparently making better time than

The Patriot, ***April 16, 1900.***

WORK WELL UNDERWAY

Workmen Hustling on Both Sides of Big River Bridge.

LABORERS ARE ORDERLY

Although the preparations for the construction of the big bridge of the Pennsylvania railroad across the Susquehanna river at Rockville are progressing rapidly, there is little as yet to betoken to an outside spectator that the building of the viaduct has been commenced.

Workmen are still engaged in digging for the foundations of the abutments on both sides of the river and a number of houses have been removed from the path of the bridge on both sides. The workmen on the west shore are still in the lead and have their abutment foundation well under way. There the work of preparation for the

The Patriot, ***May 2, 1900.***

above, ***The construction-railroad trestle has been extended into the river from the west side, preparing the way for the building of piers.***

right, ***Explosives were used to lower the river bottom both for pier construction and, as seen here, to provide enough draft in the shallow water to allow the contractors' flatboats to maneuver.***

above, ***Construction of Pier 47 has risen to the fourth course of stone. The hollow area where the workmen are standing will be filled with concrete. In the background, the west abutment wall is beginning to rise.***

above, ***As an eastbound PRR passenger train rolls across the iron bridge, a steam-powered derrick lifts a block of stone for a pier into the area where a cofferdam has been built to shield the worksite from water.***

right, ***Pier 45 on the Marysville side is nearing completion.***

above, ***a flatboat derrick, with its coal supply piled on its deck, is anchored east of a completed pier.***

right, ***the angled upstream breakwater is prominent in this view of a pier that has risen to the level of the springing line (the base of the arches).***

above, ***Piers on the west side stand in various stages of completion. Note that the three nearest the camera, Piers 45, 46, and 47, have been built longer than the others to accommodate the main line's curve northward toward Marysville.***

left, ***The first wooden arch center that would be used to lay the voussoirs, or rings of arch stone, was placed on July 30, 1900, between Piers 46 and 47 on the west side. Each one weighed seven tons.***
right, ***Centers have been placed to allow the building of eight arches on the west side. Work has advanced on the south wing wall of the west abutment.***

left, ***Kerbaugh workmen guide a stone block into position at the junction of the west abutment and what will become Arch 48.*** right, ***The falsework center between Pier 47 and the west abutment contains only half an arch on the nearest row of trusses because of the curving masonry that will be needed to support the curve of the mainline tracks.***

left, ***Falsework centers for eight arches are in place on the west side.*** right, ***Steam is up on one of the traveler cranes used to move stones for laying of the arches. In the background, the north wing wall of the abutment has been completed, but the south wing wall is still under construction.***

left, ***Construction of nearly all of the piers for the entire bridge is complete, with falsework arch centers rising on both the Rockville (in distance) and Marysville sides.*** right, ***Drake & Stratton traveler cranes are placing the stone voussoirs (arch ring segments) to complete Arches 4 through 8 and filling in the spandrel walls above regular piers 3 through 7 and abutment pier 8.*** right, collection of Rodney DiPaolo

text continued from page 38
could be started below river level. Cofferdams were built around the site of each pier, water was pumped out, and a shallow layer of gravel was removed. Hydraulic Portland cement was prepared and poured to form the concrete foundation for each pier.[42]

Work moved quickly, and as a result, both contractors began to hire more men. Some of the disorderly laborers who had been fired by Kerbaugh were hired by Drake & Stratton, and *The Patriot* reported that "they have been having 'good times' almost every night."[43] By June, contractors had added a night shift, with work illuminated by electric spotlights, the current for which was supplied by the Marysville Electric Light Works.[44]

The project was in full swing by July, with 800 men employed on the bridge, several hundred of them stone masons. Favorable river conditions continued to aid the work. By the last week of that month, the contractors had pushed construction well beyond their initial contract terms of eight arches on each side. On the east side, Drake & Stratton had built 13 piers, and on the west side, Kerbaugh had built 16 piers and was preparing to start the first arch.[45] Just a week later, the last pier was started, and the first wooden falsework arch centers were put in place.[46]

Built of four-inch yellow pine and bolted together with iron fastenings, these arch centers were immense, each weighing about seven tons. The procedure for building an arch was this: Half the width of each arch was constructed at one time. A wooden arch center was positioned between two piers, often if not always on the upstream side. Rows of 46 voussoirs, each 42 inches high, were assembled in position on these forms. The installation of the highest row of voussoirs – the keystones, each measuring 48 inches high – completed the structural integrity of that portion of the arch, and the temporary wooden centers were then moved elsewhere. Later, the other half of the arch width was completed.

With the summer weather, the bridge became a major tourist attraction, and crowds of sightseers came up from Harrisburg, especially on Sunday afternoons. In August, the *Telegraph* noted that crowds were visiting the site on both sides of the river on Sundays, when little work was going on. Many people walked out on the narrow-gauge construction-railroad trestles to get a closer look at the piers.[47]

As for the old bridge, with at least two construction seasons anticipated to complete the new span, PRR was doing its best to hold the iron bridge together. All trains were ordered to run slowly over it, and heavy passenger trains virtually crept across.[48]

First casualty

On August 1, the first fatality on the project occurred when an Italian workman, Angelo Mucci, did not hear an approaching dinky, or steam engine on the narrow-gauge construction railroad, and was run over. A few days earlier, he had sent his earnings to his wife and three children in southern Italy. Two days after the accident, a funeral service was held at St. Patrick's Pro-Cathedral in Harrisburg, and Mucci's body was buried in a Catholic cemetery.[49] Mucci was the first of six workers to die during the course of bridge construction, including two water boys, one of them 12 years old and the other 13 (see appendix).

Favorable conditions and continued low water speeded the work along, but on August 21, weather halted operations for the first time when a heavy rainstorm hit the area. By that time, three wooden arch centers had been erected on the Rockville side and five on the Marysville side.[50]

By late summer, enough of the pier masonry and falsework arch centers had been completed that the contractors began to transfer some of their laborers and machinery to other projects, leaving mostly stone masons at the bridge. Drake & Stratton moved some of its dinky steam locomotives to western Pennsylvania and some of its workers to Spruce Creek, where the firm was starting to rehabilitate the original 1,147-foot-long tunnel, which dated from 1850.[51] At that time, Drake & Stratton had 1,000 employees working at various points on PRR lines, and Kerbaugh had the same number just on PRR's Pittsburgh Division.[52]

With the piers far along, Kerbaugh began to dismantle the narrow-gauge construction trestle and prepare traveler cranes that would ride on rails above the bridge superstructure once the arches were begun. At the same time, another temporary narrow-gauge track was laid on the south half of the completed piers. Carloads of sandstone continued to arrive at the site, but the project swallowed them up as quickly as the masons could prepare the blocks. On September 5, the *Telegraph* reported: "Stone is arriving in large quantities and no further delays are anticipated now."[53] Just three weeks later, on September 26, the *Star-Independent* noted that eight wooden falsework arch centers were in place on each side of the river, but "the work of constructing the stone arches has not begun as yet, on account of the lack of stone."[54] A week later, the same paper noted: "There is very little progress being made on the Rockville Bridge for the past few weeks on account of the lack of cut stone which are being shipped from the quarries at Johnstown."[55]

A minor ripple in labor relations occurred in October, when H.W. Staley, a boarding-house keeper, had a foreman of construction, John Pendergast, arrested and jailed on a charge of slander, which "caused a suspension of work on the bridge."[56] Apparently his release was secured and the trouble settled, because work was reported as going on normally just a few days later.

Shipments of stone resumed, and mild fall weather allowed the contractors to advance the work. By late November, two arches on the Rockville side were completed.[57]

Layoffs were evident just after New Year's Day 1901, and by mid-January, work was suspended for the season. At that point, all 47 piers were completed to the height of the skewbacks, or triangular-shaped stone blocks where the arches meet the piers, and eight half-width arches (25 feet) had been completed on each side. The project went into hibernation for the winter, but passengers riding trains over the old bridge still gaped at the sight. And with the December opening of the Harrisburg Traction Co. trolley line to within walking distance of Rockville, hundreds of sightseers still visited each Sunday. The first construction season had been successful for the contractors, because "altogether they did not lose 10 days by reason of bad weather or high water."[58] ★

RJ Tower

Front and rear views show RJ block and interlocking tower shortly after its completion in 1898. Both, collection of Eric Ready

During the late 1890s, the Pennsylvania Railroad was starting to build up its main line to handle even more train traffic, adding a third, and, in places, a fourth track. One indication of this appeared at Rockville in July 1898, when PRR built a new two-story brick and frame block station and telegraph office designated RJ (Rockville Junction) to replace the previous RJ office. Often known as interlocking towers, these buildings housed an "interlocking machine" that was designed for safety. It mechanically prevented a block operator from inadvertently giving clear signals to opposing trains.[1] The machine at RJ boasted 104 levers to control the maze of switches and signals governing the trackage at the east end of the bridge, where the PRR main line, Northern Central-operated Rockville Branch, and Schuylkill & Susquehanna railroad came together at the entrance to PRR's Harrisburg Yard. The tower was put into service on Sunday, September 11, 1898.[2]

Depending on the route taken, levermen might have to throw as many as eight levers to pass a single train through the junction. At one point three levermen at the tower decided to keep track of the total number of times they threw levers in one 24-hour period – their tally was an amazing 18,026, giving some clue as to how busy this point really was. Not quite so busy was the tower at the west end of the bridge – known by the telegraphic symbol PR, for Perry County – which governed fewer switches and signals.

One of the operators at RJ Tower, H.F. Rupley, kept a log of the number of trains that passed that point in a year's time, ending September 30, 1900. He tallied a grand total of 76,405 trains, making an average of 209 trains per day.[3]

The mechanical interlocking machine at RJ Tower (later known as Rockville Tower) lasted until November 23, 1942, when it was replaced with a Union Switch & Signal electro-pneumatic machine having 13 levers controlling a total of 40 signals, 15 levers controlling 23 switches, and one "traffic" lever, for a total of 29 working levers. Nine spare levers and five spare spaces for levers gave it the capacity of a 43-lever machine.[4] ★

At the middle of the river lay the dividing line between the two contractors' work. Drake & Stratton constructed Arches 1 to 24 and Kerbaugh constructed Arches 25 to 48. In this view looking east, a group of employees poses in front of the still-unbuilt Arch 26, which lies between Piers 25 and 26. The two separate sets of traveler crane rails, indicating the limit of each company's work, can be seen above the next pier to the right of the men; that is Pier 24, an abutment pier – and the exact center of the bridge.

5. On to completion: 1901-1902

The piers of the stone Rockville Bridge lay unfinished for 10 weeks in the winter of 1901 while the contractors re-deployed their forces elsewhere. The newspapers of the day didn't explain why work couldn't continue, but one contributing factor was likely the primitive nature of the technology of concrete, which filled the center of each of the 47 piers and the area between the spandrel walls that rose above each pier and extended over the outside edges of each arch. Once cement is poured, it needs to be warm enough to cure properly as it hardens into concrete; otherwise, it loses strength. Technology has since eliminated or reduced many of these limitations, but in 1900, that wasn't the case.

Physical work may have lain dormant, but PRR was moving ahead with plans involving the bridge in other ways. To accommodate a realignment of the main track, the railroad had earlier started buying properties for a new right-of-way. By January 1901, the company had acquired 45 tracts of land "by paying good, stiff prices" and was anticipating slicing away the end of Blue Mountain south of town, at the west end of the stone-arch bridge.[1] In addition to the existing track alignment, which ran between the river's west bank and the Northern Central Railway tracks and yard, the new alignment would cross the NCR on a short bridge just west of Rockville Bridge, then proceed north through town on the west side of the NCR yard before rejoining the original PRR main line at NC Tower, north of town. (NC was so named because it once guarded the grade crossing of the PRR main line and the Northern Central main line at the east end of that railroad's Dauphin-Marysville bridge.)

Although PRR didn't reveal its intentions at that time, the railroad was planning to build a new Marysville passenger station along the new route, and excavate a $60,000 public underpass beneath the Marysville yards to eliminate three deadly grade crossings. This was what became known as the Marysville subway, the term "subway" in those days having a connotation of a public-thoroughfare underpass, in addition to that of an underground mass-transit railway.[2] At the same time, PRR had already started work on the Market Street Subway in Harrisburg and would soon begin the Herr Street Subway, also in Harrisburg. Further, the railroad was about to begin a major expansion and realignment of its main line at the nearby towns of Cove and Duncannon. At Duncannon, the work would transform the main line from two tracks through the crowed heart of town to four high-speed tracks on a new embankment along the Susquehanna.

Meanwhile, internal PRR memos were circulating regarding the placement of telegraph lines on the new bridge. Rather than erecting unsightly poles, the company planned to lay the wires as an underground conduit placed in wooden pipe buried in a passageway, surrounded by ballast and centered between the middle two tracks.[3]

"This is the longest bridge of its kind in the world. There is none in existence anywhere near its length. So expensive is the work that it has not been the practice to build stone arch bridges of more than eight or 10 spans, but the Pennsylvania Railroad when it wants a good thing does not stop at the expense."
– Harrisburg Telegraph, August 12, 1901

Work resumes

It wasn't until late March 1901 that any direct work resumed on the bridge. The vacant boarding houses were being prepared for reopening, and a small crew of workers was starting to fabricate more wooden falsework arch centers. A setback occurred on March 25 when four arch centers dropped into the river without warning, giving some of the workers narrow escapes.[4]

The coming of the spring freshets brought a new hazard for

boatmen on the river. Each spring, lumbermen from the upper Susquehanna assembled huge masses of timber cut from the forests north of Williamsport, floating them downriver to market as "rafts." With all 47 piers of the stone-arch bridge in place, the raftsmen found it rough steering to negotiate their vessels between the piers of the iron bridge and then immediately through those of the stone-arch bridge. In addition, the stone piers provided only 70-foot openings, compared to the 160-foot openings between piers of the iron bridge.[5] Two raftsmen called out Andy Simmers of Harrisburg, a 40-year veteran of rafting, to Clarks Ferry, just above Duncannon, asking him to pilot them because they were sick. Their rafts together contained 19,000 cubic feet of oak. Simmers quickly figured out that the captains weren't sick, but instead were afraid to steer through the Rockville piers. He took the job anyway, and brought the lumber to Harrisburg, where it tied up before continuing to its intended destination at Marietta, another 20 miles downriver.[6]

The resumption of work on the bridge initially put about 100 men on the project, including those working on the shore, on the river, and as stone masons.[7] At the same time, fair weather was returning, bringing out more crowds of Sunday sightseers. The Harrisburg-to-Rockville trolleys began to run more frequently, and with larger cars.[8] "Several thousand persons were attracted to Rockville yesterday to witness progress of the work on the bridge," reported the *Star-Independent* on a Monday in June. "The trolley cars run every 15 minutes and every car was well loaded with passengers. A few restaurants have been established at Rockville since the trolley line has been opened to that point."[9] Actually, the trolley tracks did not yet reach all the way to Rockville, but did bring passengers to within sight of the bridge and from there, they could walk the last half-mile or so.[10]

Dredging to recover coal washed downstream from collieries in the western anthracite fields was an active business, both before and long after the opening of the stone Rockville Bridge. Collection of Eric Ready

Sunset of the canal era

The scene on the Rockville side was changing rapidly. When officials drew up plans for a four-track approach to the east end of the new stone bridge, they didn't include a bridge over the Pennsylvania Canal, which PRR owned through a subsidiary. The canal had lost money since PRR took it over in 1857, but business dwindled even more as the waterway was successively cut back. In 1900, the canal company made minimal repairs to get the waterway back into shape before reopening for that season, but a large user, the PRR-owned Susquehanna Coal Co., transferred its business from the canal to the PRR. Moreover, the canal connection to tidewater at Havre de Grace, Md., had been cut, limiting traffic to local business only. As of March 1900, 259 canal boats were in use, but the company continued to run a deficit and as it turned out, that year would be the last season for active navigation.[11]

Now, in early 1901, the canal channel was being filled in with dirt for the railroad's embankment leading to the bridge, severing what in the 1830s had been a novel and historic means of transportation. Most of the remaining users were pleasure boaters, swimmers, and coal-dredge operators, who salvaged anthracite silt out of the river that had washed downriver from collieries around Lykens and Shamokin. They scooped it out, dried it, and sold it cheaper than conventional coal dealers. One newspaper fretted that the end of the canal would bring the end of the dredging,[12] but in fact it continued until well into the 1940s. By summer 1901, the canal would be shut down and employees laid off or transferred to PRR work, with a reporter noting that boats had not been seen for three months. The railroad was beginning to fill in the channel at several locations, and the *Telegraph* wrote: "The Pennsylvania Canal is nearly a thing of the past as far as Harrisburg is concerned. It is nothing more than a big ditch .... Down town the banks are being encroached upon and the beginning of the end is apparent at more than one place."[13]

Excavation at Marysville

On the west side of the river, H.S. Kerbaugh had returned from a vacation in Bermuda, "where he spent a short time to get away from derricks and steam dirt shovels,"[14] to oversee the resumption of activity on the bridge, and to start grading the new line at Marysville and building the Marysville subway. For the new alignment, an immense amount of excavation was needed, which would result in "a wonderful ripping up of the earth and a tearing down of the mountain nose and sides … Between the bridge and Marysville, it was necessary to shave away the face of the mountain, and dynamite, supplemented by the work of an army of Italians, Hungarians, and Polanders, did the work with neatness and dispatch, so that the road bed is slowing assuming shape …." Using the same grade of sandstone as that in the Rockville Bridge, Kerbaugh's men were also building a single-arch stone bridge over Fishing Creek at Marysville.[15] The re-alignment of tracks also meant the removal of the 36-stall Northern Central Railway roundhouse.

These were only a few of Kerbaugh's projects with the Pennsylvania Railroad. He was also building a new PRR passenger station at Pittsburgh, and held additional contracts for grading on the railroad's Pittsburgh Division (Altoona-Pittsburgh) and Middle Division (Harrisburg-Altoona), including the stretch from Cove to Sherman's Creek, near Duncannon. On this stretch, railroad officials envisioned, but did not carry out, a plan to straighten the main line even more by tunneling under the ridge just south of Duncannon.[16] In all, Kerbaugh employed 3,000 workmen at the time,[17] about 400 to 500 of them at Marysville alone.[18]

Upriver at Duncannon, PRR had given the mainline realignment project to Drake & Stratton, which put 350 men on the job, working on both day and night shifts.[19] The railroad added "dirt trains," which shuttled fill removed from the Marysville excavation to the grading needed to build the new four-track main line at Duncannon.[20] With all of this labor going on, both towns experienced an extreme housing shortage. "Not an empty house can be found in [Marysville]," reported the *Telegraph.* "A number of families that have been crowded out are living in one or two rooms."[21] At Duncannon, the same paper said, "a number of people are compelled to double up …. At least from 20 to 30 houses could be rented on sight, but not an empty house can be found anywhere"[22]

Life around the work camps was not nearly as wild as it had been at the Spruce Creek Tunnel job, but still turned lively at times. A judge sentenced one Philip Johnson to a year in a penitentiary for assaulting a C.D. Dobs, a boss on the bridge project.[23] On another occasion, someone tried to sabotage the narrow-gauge construction railroad that was used to haul concrete out onto the

top, ***An H.S. Kerbaugh steam shovel crew begins to chew away the base of Blue Mountain on the Marysville side of Rockville Bridge.*** right, ***Some of the men who worked at the job of excavating the mountain were Marysville residents, including, second from left, Harry Kass; third from left, Earl Geib, and fifth from left, Roy Eppley.***
Both, collection of Jim Roberts

In a view looking west from Rockville, a westbound passenger train crosses the iron bridge while, just beyond, construction continues on piers of the stone-arch bridge. Collection of Eric Ready

bridge, but engineer Harry Ziegler saw the obstructions on the track and got his "dinky" engine stopped in time.[24] In what leisure time they had, some men went hunting or fishing while others drank or gambled. One noontime in June 1901, traveling Salvation Army workers stopped at the Rockville end and held a service for workers. "It was well attended and many of the men joined in the services and the singing," reported the *Telegraph.* [25]

By mid-summer, the number of arches completed or under way stood at 13 on the Rockville side, where Drake & Stratton had 130 men working, and 15 on the Marysville side, where Kerbaugh had 170 men working.[26] Progress continued though July and August, and on the afternoon of September 7, Kerbaugh's men cheered and blew the whistles of the steam dinkies and derricks to celebrate the keying, or setting, of the last keystone in the last arch on that side. The next steps were to build more ashlar walls and place concrete in the spandrel walls. Drake & Stratton had earlier completed the last of its river arches, but the terminal arch on the east side (Arch 1) overlapped the existing abutment of the iron bridge so that it couldn't be completed until part of that structure was removed.[27]

A president passes

A somber interruption to work came on September 16 and 18 when the body of assassinated President William McKinley passed the construction site twice within 48 hours. The president had been shot at the Pan-American Exposition on September 6 and lingered several days before dying. With its engine and five Pullman cars draped in black, the funeral train on Monday morning, September 16, left from Buffalo, en route to Washington. Moving southward through Williamsport, Pa., and along the east side of the Susquehanna through Sunbury and Millersburg, it passed Rockville and RJ Tower a few minutes before its 4:50 p.m. arrival in Harrisburg. Guarded by a soldier and a sailor, the flag-draped coffin was visible through the windows of the observation car in which it rode. Along the line, train movements stopped a half-hour before the special arrived. An estimated 40,000 people crowded the tracks in downtown Harrisburg alone. While the train changed crews and engines before heading over the Cumberland Valley Railroad bridge toward Washington, the crowd of several hundred at the station softly sang, a cappella, the hymn "Nearer, My God to Thee," some of the lyrics of which had been the last words spoken by McKinley before he died.[28]

A few minutes after midnight on Wednesday the 18th, another funeral train, running in three sections, passed through Harrisburg from Washington, rolling westward over the iron Rockville Bridge en route to McKinley's burial in Canton, Ohio. The first of the three trains carried newspaper reporters and minor government officials; the second carried President Theodore Roosevelt, the cabinet, McKinley's body, and members of his family; and the third section carried army and navy officers, including Admiral George Dewey, hero of the Spanish-American War.[29]

Work moves quickly

Enough masonry and concrete had been placed that the railroad took the next step of finding a supplier for sealant. On September 24, the Vulcanite Paving Co. of Philadelphia offered PRR a proposal to apply a waterproof coat of asphalt to the concrete on the bridge deck surface, and the next day, PRR awarded that firm a contract to do the work at 18 cents per square foot.[30]

Stonework was nearly complete on the bridge when the railroad's annual fall track and property inspection took place. These were highly celebrated events in which as many as five trainloads of PRR officials toured some part of the railroad, usually including the New York-Pittsburgh main line and often taking several days to do so. Coming east from Pittsburgh, the 1901 inspection party stopped at Harrisburg on October 9 and while some officials visited the yards and roundhouse, others drove carriages to Rockville to inspect the bridge, crossing to the Marysville side to view the whole project. After spending the night at Harrisburg, they continued to Philadelphia the next day,

left, ***The difference between the breakwater of a regular pier, right, and double-width abutment pier, left, is apparent in this view. The added strength of the abutment pier provides a convenient stopping place for arch construction.*** right, ***The cross-shaped cross-section of the falsework arch centers is noticeable in this view on the west side.***

left, ***After the construction season of 1900 ended, each contractor had built eight arches, as seen in this view of Kerbaugh's work on the Marysville side.*** right, ***Soon after work resumed in March 1901, the building of arches continued, with at least 12 of them complete or substantially so in this view.***

left, ***The north halves of these arches are being constructed first, with an internal spandrel wall (not shown on construction diagrams) built to allow filling of that half with concrete while the south half of each arch is completed.*** right, ***A top-side view shows ashlar being lifted by a traveler crane. One of Kerbaugh's "dinky" steam engines is waiting behind the flat car from which the stone has just been lifted.***

left, ***With the north half of the arches complete, workmen finish the job by building the south half of each arch.*** right, ***Using a ladder braced in a flat boat, two men have climbed to the springing-line level of Pier 46 and are standing on a temporary wooden perch to perform finishing details at the point where Arches 46 and 47 meet.***

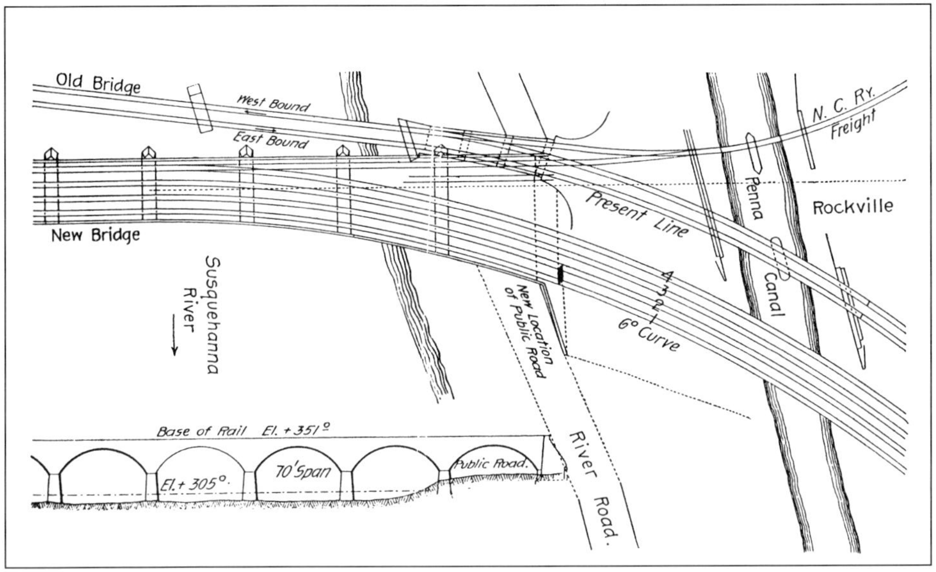

A diagram published in* The Railroad Gazette *in October 1901 shows the positions at Rockville of the iron bridge, the stone-arch bridge, the approaches to both, and the Pennsylvania Canal.

The masonry superstructure of the bridge, including the west abutment and its wing walls, is nearly complete, but the earthen fill to support the PRR main line has yet to be placed behind the abutment. Collection of Rodney DiPaolo

above, ***The masonry is nearing completion on Drake & Stratton's contract for the work on the Rockville side. In this view looking west, a covered traveler crane is positioned above Pier 8, the first abutment pier on the east side. It can be distinguished not only by its double width but also by the safety refuge offset above, which is designed to provide a place for a person to get out of the way of a train. Visible through Arches 5 and 6 is the 1877 iron bridge. To the left, two more abutment piers can be seen, each demarking a group of eight arches; the distant one is Pier 24, the middle of the bridge. Floating in the water are trusses from the wooden falsework arch centers, no longer needed.*** From Art Work of Harrisburg, Lancaster, and York, Part 4, collection of Duane Hammaker

The west approach to the new stone-arch bridge has been completed, track has been laid, and lower-quadrant semaphore signals have been installed. Those details, plus the presence of dozens of people gathered on the bridge, suggest that its opening may be imminent. Collection of John and Peggy Smith

averaging 20 mph to allow for maximum visibility, and on the succeeding day, they concluded the tour by covering the New York Division.[31]

Drake & Stratton had already started to reassign workers to the Duncannon project.[32] Less than three weeks after the PRR inspection, on October 26, Kerbaugh finished the stone work on the Marysville side, and Drake & Stratton was planning to finish its stonework the next day.[33]

Soon, workers began spreading cinders in the area between the spandrel walls and on top of the arches, which was the next layer above the asphalt that covered the concrete filling.[34] On top of this, the railroad would lay track and spread rock ballast. The job of laying asphalt continued through the end of January.[35]

A 21-foot flood in mid-December closed the Market Street Bridge in Harrisburg but not the adjacent People's Bridge. The previous high-water mark was 27 feet during the May 1889 flood. The *Telegraph* reported: The contractors in charge of the improvements at Rockville [and] Duncannon report very little damage."[36] One of the wooden arch centers floated downstream about two miles, washing up on the east shore at Coxestown. It was later towed back by a team of six mules and two horses.[37]

A few days later, on December 19, Kerbaugh finished the stone-arch work on the Marysville subway and was preparing to lay a level asphalt road in the passageway in preparation for opening it about January 1.[38]

Completing the job

By the end of the year, the night shift on the bridge had been abolished. While the heaviest kinds of work began to wind down, workers pushed forward on other tasks.[39] On the bridge deck, much of the cinder fill had been completed, and ties and rails were put down. A construction train was the first to cross on February 16, behind Engine 587, a Class H1 2-8-0-type locomotive built by Baldwin in 1882.[40] By the end of that week, three of the four tracks had been laid, and the wooden pipe in which the telegraph conduit would be placed (within a stone passageway) was being installed. A trainload of ballast arrived, and the contractors began at once to distribute the stone between the ties.[41] The fourth track was laid about March 1.[42]

All was going well in early March when the ice on the river broke up and the water began to rise again. Contrary to fears of those living along the banks, the ice did not gorge at the railroad bridges.[43] But the rising water damaged the Clarks Ferry Bridge upriver and ripped away part of the old wooden Market Street ("Camelback") Bridge in Harrisburg. After a stone block on one of the piers of the iron Rockville Bridge was knocked out, PRR officials detoured the eastbound *Atlantic Express* over the Northern Central Railway to Bridgeport (Lemoyne) and then across the Cumberland Valley bridge to Harrisburg. After inspecting the pier, they allowed service to resume. They considered tearing down the old Clarks Ferry Bridge to prevent it from washing down and damaging one or both of the Rockville bridges, but did not do so.[44]

On March 13, the *Telegraph* boldly reported that the bridge would open on Easter Sunday, a little more than two weeks away.[45]

PRR Chief Engineer Brown, who had come to town to join resident PRR engineer George Nauman in placing a proposal before Middletown Borough Council for two underpasses in that town, "paid a visit to the bridge and inspected it, expressing a desire to get it in shape as soon as possible," the *Telegraph* reported. Hundreds of men were at work seven days a week, and thousands of sightseers still flocked to the bridge on Sundays. Cinder and ballast filling continued.[46] Switch and signal connections had yet to be made, but the details were rapidly coming together.[47]

The first train

Finally, the bridge was turned over to the Middle Division, the operating department having jurisdiction over the 130 miles of main line between Harrisburg and Altoona. A newspaper reported that all of the track men between Harrisburg and Newport (a distance of 25 miles) would be on hand to make the track connections.[48] At about daybreak on Sunday, March 30, track workers cut the main line at each end of the iron bridge and bolted the rails to those leading across the stone-arch bridge. Silas S. Kuhn of Harrisburg was an assistant track foreman at the time[49] and helped make the transition.[50]

The first train to cross, about 9 a.m., was an eastbound passenger train. It was No. 20, the *Atlantic Express*, a daily run that was the first eastbound long-haul passenger train to arrive in Harrisburg each morning. Scheduled to leave Pittsburgh each day about 3 a.m., the train ended its run in Jersey City, N.J., across the Hudson River from New York City, at 2:40 p.m.

Hauling the train was engine No. 30, a Class D16a 4-4-0 type fast passenger locomotive that had been built in 1897 at PRR's Juniata shops in Altoona. It was one of 429 D16 engines built at Altoona for PRR and its subsidiaries between 1895 and 1905. At the throttle was engineer William Kauffman, and the fireman was H.F. Stewart. In charge of the train was conductor John P. McPherran, and the brakemen were William Young and Richard Lightner. Passengers crowded the windows, waving to the scores of railroaders and spectators who had gathered for the event.[51] In the end, the bridge had cost $975,150, very close to the $1 million estimated at the start of construction.[52]

Together with the cost of excavation on the approaches and yard at Marysville, the total ran well into seven figures. To people in Harrisburg, Rockville, and Marysville, it sounded like an impressive sum, but it was loose pocket change to the Pennsylvania Railroad, which had just spent $110 million to buy stock control of several competing trunk-line railroads and within a few years would spend $112 million more to build a system consisting of tunnels, yards, and a palatial new station in New York City.

Fifty years after the fact, a group of retired track workers – Kuhn at age 101, Al Hilner, and Charles and Al Wade, all of Harrisburg – recalled the opening day for Harrisburg *Evening News* reporter Al Hamman. He wrote that they "supplied a [previously] unpublished note … saying the heavy rain caused the roadbed to settle, making the track uneven as the train passed over. The supervisor, J.E. Niemyer, was so disturbed … that he 'threw his hat in the river.' Charley Wade said he worked until 10 p.m. that Easter Day because of the track trouble. His brother and Hilner said they and many more worked all day and all night under 'Sandy' Anderson, boss of the floating gang."[53] ★

The first train to cross the stone-arch Rockville Bridge was the eastbound* Atlantic Express, *Train 20, on March 30, 1902. Pulling it is PRR Engine 30, a D16a-class 4-4-0-type locomotive.
Collection of author

"At Rockville, just above the capital city, they have thrown across the Susquehanna a four-track bridge of monolithic stone seven-eighths of a mile long and stepped in graceful arches as enduring as the mountains that look down on the beautiful river. Bridges, like men, have their tables of mortality; but in the expectancy of life accorded to American bridges here is a structure to which no limit of years may be assigned; it has been built to last forever."

– Frank H. Spearman,
The Strategy of Great Railroads, 1904

Costing $975,150, the stone-arch Rockville Bridge is complete and ready for service in the spring of 1902. With unused sandstone still stacked in neat piles, the Drake & Stratton worksite, located inside the mainline curve at Rockville, appears to be nearly abandoned. In the foreground, a maze of complicated junction trackage includes several double slip switches to allow maximum flexibility in moving trains between PRR mainline tracks or, to the right, to or from tracks of the PRR Rockville Branch or the Reading's Schuylkill & Susquehanna Branch.
Collection of Rodney DiPaolo

6. Open for business

Crew members who worked aboard trains that crossed the new bridge on its first day boasted proudly of that fact. Freight enginemen, however, began to fret that the sharper curves at both ends of the new bridge meant increased friction on their heavy trains, making for a harder pull. But the change must not have been too extreme, because no evidence points to an epidemic of stalled trains.[1]

Some finishing work on the bridge remained, but PRR was expanding so rapidly that officials immediately began to turn most of their attention elsewhere. Kerbaugh was still building a new passenger station at Pittsburgh, won a PRR contract to build a new freight classification yard in Altoona,[2] and would soon get even more work. Drake & Stratton was still working on the new mainline alignment at Duncannon. PRR engineer George Nauman began to review plans for remodeling and expanding the Harrisburg passenger station,[3] and continued to oversee the building of new subways at Harrisburg and nearby Middletown. Also, in conjunction with the realignment of the main line, new passenger stations were being built at both Marysville and Duncannon.

The tracks opened at the west end of the new bridge curved northward immediately, running between the river and the Northern Central Railway main line and yard. Construction continued on what would become a double-track passenger line through Marysville west of the NCR facilities, which would lead past the site of the new depot. As part of the construction of this track, crews built sandstone retaining walls and abutments to support a steel girder bridge over the Northern Central main line that entered Marysville Yard from Lemoyne and Enola.

On the east side, three weeks after the stone bridge opened, PRR began the task of removing the old abutment at Rockville, where the two bridges intersected. It was necessary to remove the three 25-foot-wide stone arches leading to the iron bridge and most of the old east abutment in order to complete the terminal arch (Arch 1, the easterly of the 48 arches) and east abutment for the stone bridge.[4] The work of demolishing the three arches was under way by mid-April,[5] and the foundation for the new arch was laid on May 28.[6] By that time, PRR had already begun pulling up the rails on the iron bridge.[7]

One of the first freight trains, if not the first, to cross the bridge was this westbound movement hauled by three-year-old F1a-class 2-6-0-type engine No. 650, captured by photographer W.N. Jennings. It is carrying green flags, indicating an extra movement.
Collection of Rodney DiPaolo

Having no more need for the iron bridge, PRR sold it to Henry Levi & Son of Philadelphia, who contracted with John B. Fraim of Philadelphia to remove it.[8] Either Levi or Fraim made a proposition to sell the iron spans to the Harrisburg Bridge Co. to replace the public toll bridge at Harrisburg – the wooden swayback Camelback Bridge that was damaged in the March 1902 flood.[9] The bridge company directors chose not to accept the offer, because they subsequently hired Kerbaugh to build a new steel-girder bridge on masonry piers.[10]

Debut of the *Pennsylvania Special*

The train that would become the Pennsylvania Railroad's flagship luxury flyer made its inaugural overnight trips between New York (Jersey City, N.J.) and Chicago on June 15, just 2-1/2 months after the new Rockville Bridge was completed. That was the extra-fare *Pennsylvania Special*, which in 1912 was renamed *Broadway Limited.* PRR was locked in a speed war with its rival, the New York Central & Hudson River/Lake Shore & Michigan Southern system, for long-haul passenger service. On the same day, both railroads inaugurated 20-hour trains – PRR's was the *Special*, and NYC's was the *20th Century Limited.* Up to this time, trains took from 23 to 27 hours to make the run, which covered 912 miles via the PRR route. On the first trip west, the four-car *Special* arrived in Harrisburg on time at 5:50 p.m., changed engines and crews, and took off like a shot behind Engine 443 with engineman Calvin C. Miller at the throttle. After slowing for the curves at each end of Rockville Bridge, Miller accelerated the train again, passing the four miles between Cove and Duncannon in 2 minutes 31 seconds, or an average of 95 mph. After several more changes of engine and crew, the train reached Chicago the next morning three minutes ahead of schedule, as did its rival, the *20th Century Limited.*[11]

In freight traffic as well as in passenger service, the railroad was booming. On the Middle Division and in the Harrisburg Yard, PRR employed a record 678 engine-service employees – 254 road enginemen and 265 road firemen and 75 yard enginemen and 84 yard firemen.[12]

opposite page, ***It is August 1902, and dismantling crews have removed the trusses of Spans 1 and 2 on the 1877 iron bridge at Rockville. A crane at lower right has loaded a plate-girder span that once carried the double-track main line across the Pennsylvania Canal onto a flat car, and is in the process of lifting a second span.*** Railroad Museum of Pennsylvania (PHMC)

Contractor John Fraim has advanced westward in removing trusses of the iron bridge, leaving only eight of the original 23 remaining. Fraim suspended work for the winter of 1902-03, leaving the bridge with eight spans standing from December 1902 through March 1903. Leib Image Archives

Meanwhile, projects related to Rockville Bridge and its approaches were continuing. PRR assistant engineer George Nauman was overseeing the straightening and smoothing of the main line from Cove to Sherman's Creek, and the construction of the new four-track line along the river at Duncannon, the first phase of which was opened on July 25. Nauman was aboard a special train arranged by W.B. McCaleb, Middle Division superintendent. Other passengers on that train were Harrisburg Mayor Vance McCormick and Lyman Gilbert, the Harrisburg lawyer with whom George Nauman's brother Spencer G. Nauman Sr. would later affiliate.[13] Taken together, the work at Rockville, Marysville, and Duncannon was a huge investment by the Pennsylvania Railroad, representing, in the words of a *Telegraph* reporter, "the expenditure of millions of dollars .... The people of Harrisburg have no idea of the vast work that is being done within 20 miles of this city, but it is a stupendous enterprise and can be charac-

right, ***Construction of a new double-track passenger main line into Marysville is under way in the foreground of this view by W.N. Jennings during the late summer of 1902, with placement of steel girders above the Northern Central tracks. In the interim, all trains, freight, as shown, and passenger, are using the PRR tracks that lie closer to the river. The passenger main line was opened in December 1902.***
The Library Company of Philadelphia

"With the driving of the last spike and the connection of the last switch the great stone bridge at Rockville will be formally opened to the traffic of the Pennsylvania Railroad on Sunday. This will mark the completion of a great undertaking. It is one of the wonders of the world and is a conspicuous example of the modern methods of railroad construction. It is worthy the giant corporation which is building deep and enduring foundations for its colossal business."

– Harrisburg Telegraph, March 28, 1902

At the west end of Rockville Bridge, a westbound freight train with five cabin cars (cabooses) passes the junction where the passenger main line, which passes over the steel girder bridge, crosses over several freight tracks that connect Marysville and Enola Yard.
Collection of Ted Xaras

terized by the one word – wonderful."[14] Other railroad work, closer to Harrisburg, was also going ahead. PRR was planning to enlarge its Harrisburg Yard, and Drake & Stratton was given a contract to eliminate a grade crossing in the city by building a subway at Herr Street, estimated to cost $90,000.[15]

The job of removing the iron Rockville Bridge got under way by mid-summer, and the first of the 23 spans was gone by August 15.[16] Two months later, contractor Fraim had removed the sixth span and was anticipating going at the rate of one per week.[17] By late November, 60 men were on the job and they were removing the 12th span. A month later, they had removed 15, leaving eight still standing. At this point, Fraim suspended work for the winter.[18] His workers resumed the dismantling job when the weather improved, starting in March 1903.[19]

Beginnings of Enola Yard

Meanwhile, PRR had still bigger plans in mind that would reshape the landscape of the Harrisburg-area railroad scene. On November 20, 1902, the Pennsylvania awarded contracts to H.S. Kerbaugh to build a massive, four-mile-long freight classification yard at Fairview, or Enola. The job didn't stop there – the work also included grading right-of-way and laying track for a new low-grade freight line along the west shore of the Susquehanna to Columbia, with a 2,209-foot-long stone arch bridge over the Susquehanna at Shock's Mills, 20 miles south of Harrisburg.[20] This was part of PRR President Alexander J. Cassatt's strategy to avoid congestion by building easier-grade bypasses to keep slower freight trains separated from faster passenger trains.

Within a little more than two months, 3,000 men were at work excavating the ground for Enola – more than three times as many as had worked on Rockville Bridge at its peak.[21] By April 1903, the number had jumped to 5,000.[22] The job would last for two years – same as Rockville Bridge – and would involve moving 10 million cubic yards of earth to build what would become the largest railroad yard in the world.

PRR had left space in the bridge abutment design at the west end to allow laying a track southward to directly connect with the main line of its subsidiary Northern Central Railway. Now, with plans for construction of a huge yard at Enola, it would seem to be even more important to open that connection, but for reasons that remain unclear, PRR decided against completing that link.

As a result, NCR's freights between Baltimore and Sunbury

above, ***After removing the iron trusses, contractors' employees built a narrow-gauge construction railroad trestle into the river, similar to those of the builders of the stone-arch bridge, to facilitate dismantling of the stone piers from the old bridge.*** below, ***Forty-seven men who normally wouldn't be dressed up at a construction site have donned their finery for a group photograph. The sign behind them reads: "Steam shovel men employed at the Fairview Yard, H.S. Kerbaugh, Inc., contractor." Fairview was the name used before PRR adopted the term Enola Yard.*** Both photos, collection of Jim Roberts

still had to go into Marysville yard and change direction. Now, with the prospect of more tracks being laid and more trains moving in and out of Enola, questions about the exact form of the junction trackage came into sharper focus. Correspondence among Chief Engineer Brown, Assistant Engineer Nauman, and Middle Division Superintendent McCaleb outlined the situation.

Brown wrote to Nauman on December 8, 1902: "I do not

Between Marysville station and the west approach to the bridge, the passenger main line passed over Fishing Creek bridge, which Kerbaugh employees constructed in the same manner and with the same materials as those with which they built Rockville Bridge. Collection of Jim Roberts

A quarry situated in a nook of Blue Mountain above the west end of Rockville Bridge was operated by J.P. Dyer of Pottstown, Pa. Railroad hopper cars were placed at the crusher via a siding that led off the Northern Central Railway line. Used to supply ballast for the construction of Enola Yard (1903-1905), the operation was reportedly closed by about 1910. Collection of Jim Roberts

think it will be necessary to keep up the connection with the south "Y" off the stone bridge, as our people have about given up the idea of laying the track on this "Y" …." He asked Nauman to estimate the cost of reducing the NCR grade going south into Enola, but Nauman wrote back the next day, saying that he didn't understand exactly what Brown meant and by what basis he wanted him to figure the proposed regrading.[23] What Brown had in mind was raising the NCR track to ease or smooth the grade leading into Enola, but to do that would require raising the bridge that carried the as-yet-unfinished PRR passenger main line over it.

A few days later, on December 12, which was a Friday, Brown wrote to McCaleb: "Do you intend to open the new passenger line at Marysville on Sunday? If you can postpone it for a week or two I think it would be a good thing to do for if the plan for Fairview Yards is changed it may be necessary to raise the bridge over the N.C. Ry. and we should do this before we open it for service."[24] After he sent that message, Brown wrote again the same day: "Please cancel our message … and go ahead and open the new line at Marysville when you are ready to do so. There will be no change in the passenger track at the west end of the Rockville Bridge."[25]

A proposed loop, and another bridge

Apparently, some PRR civil engineers still hadn't given up the idea of making a direct PRR-Northern Central connection, because the *Altoona Tribune* reported on a proposal to build a loop that would circle under the first arch. Because of the tightness of the curve and the limited space in which to make the turn, it was deemed impractical. Even though the plan was reconfigured to ease the curve,[26] the idea never got off the drawing table.

Another idea that surfaced in the midst of all this seemed to carry more weight than the loop proposal. In the winter of 1903, a corps of PRR civil engineers was sent into the field to run surveying lines for a proposed new Dauphin-Marysville freight-only bridge that would basically recreate the Northern Central bridge that had stood from 1858 to 1883. In addition, PRR real-estate specialists were knocking on doors around the north end of Dauphin to inquire about buying property.

The proposed new structure would be a double-track steel bridge that would cross the Susquehanna at a 45-degree angle. "A stone bridge will not be built because of the great cost of such a structure and the time required for its construction," reported the *Telegraph.* "The new bridge will be about the same length as the stone bridge at Rockville."[27] Coal trains from the north, which caused congestion by competing with the flow of east-west main-line trains using Rockville Bridge, would travel on this new crossing directly into Enola. In the end, neither the cloverleaf-like loop at the west end of the bridge nor the proposed Dauphin-Marysville new bridge was built. Eventually, PRR did build the direct wye connection that it envisioned at the west end of Rockville Bridge, but not for almost 37 years.

In February 1903 and again in April, PRR President Cassatt visited Harrisburg and Enola to personally inspect the progress on his program. In 1905, both Enola Yard and the low-grade freight line were opened.

New landmark, and a political note

When the stone Rockville Bridge was opened, it became an instant landmark, both as an engineering accomplishment and as a public symbol of the Pennsylvania Railroad. As such, it began to appear on many postcards, and The National Geographic magazine later featured a photo of it to accompany a travel piece. In service, it quickly faded into the infrastructure of the Middle Division, and its responsibility was transferred from construction crews to maintenance-of-way crews.

In the fall of 1903, the bridge took on an unusual role during a heated local political campaign. Encouraged by the climate of a national Progressive-era push for better government, a grass-roots movement emerged to oppose corrupt machine politics that had dominated Dauphin County government. Among those supporting this cause was Vance McCormick, mayor of Harrisburg and publisher of *The Patriot.* A prominent Democrat, he was a leader in the City Beautiful movement that brought city planning, parks, a pure water supply, and sanitation to Harrisburg, all elements of Progressive reform. With elections approaching in November, leaders of the Anti-Machine group held rallies in Harrisburg and around the county. The officeholders they opposed were Republicans, but the issue was more about how they governed than about partisanship – county tax rates had shot up by 25 percent in the previous three years.

On the evening of October 21, 1903, four such meetings were held simultaneously around the county, one of them at Rockville Bridge. While other communities had suitable meeting places for a large group, Rockville didn't, so promoters simply set up a platform and gas lights under the first arch. The speakers, ac-

The Pennsylvania Railroad employed horticulture specialists to beautify the grounds around major stations and other high-profile sites. One of these locations was Rockville; visible in this postcard view are ornamental shrubs, a stairway and sidewalk, and a fountain. Collection of Eric Ready

RJ Tower guards the maze of switches and semaphore signals that make up the junction between the PRR main line, center, the PRR Rockville Branch, foreground, and the Reading Railroad's S&S Branch, the track farthest right. Visible through the signal bridge is the passenger shelter for riders boarding or alighting eastbound trains at Rockville. Collection of author

above, ***A westbound late-afternoon passenger train, perhaps the*** **Pennsylvania Special,** ***crosses the west end and is about to enter Marysville behind an E2/E3 class 4-4-2-type engine, typical fast passenger power of that day. Passing through the 48th arch is the trolley track completed by the West Fairview & Marysville Electric Street Railway in October 1902.*** Collection of Duane Hammaker

left, ***When the large PR tower at the west end of the iron bridge was demolished, PRR built a modest wood-frame structure, perched at the west end of the stone-arch bridge, to replace it. Levers in the tower moved the pipes in the foreground to control signals and switches.*** Collection of Walt Ready

companied by the Commonwealth Band, arrived by trolley about 8 p.m. and marched to the site. About 200 people attended, including – although they could not vote – many women, some of them mothers holding sleeping infants in their arms.[28] "Though the atmosphere was chilly, the audience waxed warm with enthusiasm for the Anti-Machine case," the *Star-Independent* reported. "Many had come on foot or by wagon from the surrounding country and all manifested a sincere interest …."[29] As in the rest of the nation, where the Progressive reform movement left an impact but did not survive as a separate political party, the group managed to elect one of its candidates to a county row office and then disappeared.

On the railroad, freight congestion soon became so thick that PRR discontinued the *Pennsylvania Special* luxury train that it had inaugurated with so much fanfare the previous summer. It was dropped in February 1903, and not revived again until June 1905. Still running during the interim was the *Pennsylvania Limited*, another New York-Chicago flyer that encountered an unfortunate accident at the bridge in January 1904. One of the foreign workers employed by Kerbaugh in building Enola Yard had hitched a ride on a freight train. As it approached the bridge and the place where he wanted to disembark, the man jumped, knowing that the freight was not planning to stop. Unfortunately, he landed on the next track, immediately in front of the *Limited*, which threw his body for some distance. In a sad commentary, a newspaper noted: "The man was about 30 years old and was known only by the number on the pay roll."[30]

Trains did not always make it across the bridge smoothly. In December 1904, a 79-car train loaded with coal and merchandise derailed when an air-brake pipe burst on the engine, setting the brakes into emergency application. The sudden stop derailed three cars on the bridge and tossed two more, both wooden cars, into the Susquehanna, where they "were broken into kindling wood on the sharp rocks in the river," reported one newspaper. It added: "Seven panels of the iron railing was torn from their fastenings [but] otherwise the bridge was not damaged in the least."[31]

Stone-arch design flourishes, then fades

Elsewhere on the railroad, stone-arch bridges continued to go up. Two important four-track mainline examples opened in New Jersey in 1903: At New Brunswick, a 1,428-foot-long bridge with 21 spans, and at Trenton, a 1,228-foot-long bridge with 18 spans. A 10-arch, four-track stone bridge over the Brandywine Creek at Coatesville opened on September 20, 1904.[32] As already noted, the low-grade freight branch ran over a double-track 28-arch stone-arch bridge at Shock's Mills that opened in 1905.

Chief Engineer Brown was partial to stone construction where it fit, but he also approved many important steel girder or steel truss bridges in situations where the track level was high above water; where navigation was important (neither of which applied on the Juniata or most of the Susquehanna rivers) and arches would prove to be an obstacle; or where the cost of a stone bridge was simply out of reach. Brown approved major steel bridges that were opened over the Ohio River at Pittsburgh and Wheeling, W.Va., in 1890; over the Delaware River at Frankford Junction, Pa. (near Bridesburg, northeast of Philadelphia), in 1896; over the Susquehanna River at Columbia, Pa., in 1897; over the Allegheny River at Pittsburgh in 1902; over the Monongahela River at Pittsburgh in 1903, over the Allegheny River northeast of Pittsburgh in 1904, and over the Susquehanna at Havre de Grace, Md., in 1906.

As the railroad proceeded with its major expansion and re-building, it began to run out of bridges that met the criteria for replacement with stone-arch types. At the same time, the design lost its most influential champion in the company when Brown retired at the end of February 1906 at the mandatory age of 70. What appears to be PRR's final major stone-arch bridge was built in its headquarters city of Philadelphia, where the Philadelphia-New York main line crossed the Schuylkill River. Carrying five tracks, this 11-span structure was completed in 1913-1914.

The transition to reinforced concrete, steel, or a combination of those two materials, was well under way by this time. An even greater monument than Rockville Bridge was the 995-foot-long concrete, granite, and steel-arch Hell Gate Bridge in New York City, rising 135 feet over the East River, that gave PRR direct access to New England. Work started in 1900 but progressed most rapidly from 1910 to its opening in 1917. Two major bridges along the Philadelphia-Washington main line were completed in 1913. Both were built of reinforced concrete.[33] A reinforced concrete-arch bridge carrying PRR's Schuylkill Branch over the Schuylkill River in Philadelphia opened in 1918.

In the world of bridge design, the stone-arch revival had come, risen to its highest expression in Rockville Bridge, and then just as quickly faded from the scene. ★

above, ***Another stone-arch bridge built under William H. Brown's administration was the 18-arch span at Trenton, N.J., on PRR's New York-Philadelphia main line, completed in 1903.*** below, ***Another image made by photographer William H. Rau, and reproduced in large prints and on many postcards and timetables, was this expansive view from the west side looking east.*** top, Collection of author; bottom, collection of Bennett Levin

Two trolley cars on the Marysville-Harrisburg run are visible in this panoramic view showing the west end of the bridge, its approach tracks, and the trackage leading to Enola Yard. Collection of Bill Durrwachter

7. Speed and security

After the bridge went into operation, very little changed except what ran on the four tracks on top of it. Larger engines and cars came along, many of them designed by the Pennsylvania Railroad and built in the company's shops at Altoona. At its peak, about 15,000 people worked in those shops, which covered 218 acres in 122 buildings.[1] The re-inauguration of the *Pennsylvania Special* in June 1905 again brought one of the fastest long-distance trains in America to Rockville Bridge, running westbound every evening and eastbound every morning on its overnight dash between New York and Chicago.[2]

In 1910, PRR completed its sprawling $112 million Pennsylvania Station complex in Manhattan, which included twin tunnels under the Hudson River to give the railroad direct access to the heart of New York City rather than forcing passengers to ride a ferry from Jersey City, N.J., as it had for the previous 39 years.[3] [4] The wooden passenger cars then in use were vulnerable to telescoping accidents and fire, both of which could prove to be disastrous in busy tunnels through which trains passed every few minutes. To overcome these issues, PRR designed and built a family of steel passenger cars, including coaches, diners, baggage cars, and mail cars.

At PRR's urging, the Pullman Co. likewise began to replace its wooden cars with thousands of new all-steel sleepers and parlor cars, because many Pullmans operated in PRR trains through the tunnels—in fact, the most exclusive passenger trains were all-first-class, all-Pullman runs. Because most long-haul trains that crossed Rockville Bridge either started or terminated at Penn Station, steel cars became a common sight on the bridge, and their numbers multiplied rapidly. From a total of 201 such cars in 1908, the fleet grew to 3,824 by 1918 and rose to 5,501 cars by 1928.[5] Top-of-the-line trains got the newest cars, and this included the *Pennsylvania Special*, which on November 24, 1912, was renamed *Broadway Limited*.[6] The name Broadway alluded not to the theater district of New York, but to PRR's four-track "Broad Way" main line, of which it was justly proud.

Steel cars are heavier than wooden cars, requiring larger locomotives. By 1918, PRR had begun to mass-produce the kind of passenger locomotive that would become its standard, the Class K4s 4-6-2 type, which could easily handle as many as 12 steel passenger cars. Heavier and more durable steel freight cars were

A five-car train made up of steel passenger cars behind an early K-class 4-6-2-type engine is approaching Marysville. Collection of Walt Ready

likewise coming into general use, and PRR developed new classes of freight steam engines to haul them. Among them were the Class H9s 2-8-0 type freight engine; the Class L1s 2-8-2 type, a cousin of the K4s that used an identical boiler; and the even heavier and more powerful Class I1s 2-10-0 type for slow freight and Class M1 4-8-2 type for fast freight.

Moving the division boundary

Two administrative changes took place in PRR's operating department about this time, 10 years apart. Since early days, the Rockville bridges were part of the railroad's Middle Division, which extended from Harrisburg to Altoona. On June 1, 1903, Middle Division headquarters was transferred from Harrisburg to Altoona, and the Philadelphia Division headquarters was moved from Philadelphia to Harrisburg. In conjunction with this change, PRR officials moved the boundary between the two divisions from Harrisburg (Dock Street) to Rockville, although they did consider moving it farther west, to Marysville.[7]

A lone figure poses at the base of the mast of a pair of lower-quadrant semaphore signals at the west end of Rockville Bridge. Collection of Walt Ready

Ten years later, that is exactly what happened. In 1908, Marysville Yard was closed, but PRR reopened it on August 1, 1912, when the railroad transferred its Preference Freight classification yard there from Harrisburg.[8] This brought overlapping jurisdictions where the Middle, Philadelphia, and Baltimore (because PRR's Northern Central Railway subsidiary still entered Marysville) departments came together at Marysville Yard. "With the heads of three different departments giving orders, things became complicated and a month ago, an order was issued from the office of Superintendent G. W. Creighton to have the entire Marysville yards looked after by the Philadelphia Division," reported the *Telegraph* on November 1, 1913.[9] With this directive, PRR extended the Philadelphia Division westward to NC Tower (later known as Banks tower), west of Marysville, shortening the Middle Division for a second time and placing Rockville Bridge in the Philadelphia Division. Another change took place in 1914, although it was mostly a financial transaction. PRR had held majority control of Northern Central and managed it as a subsidiary, but on July 29, the parent company took full control with a 999-year lease.

Earlier that same year, an unusual fatality took place on the bridge. Many fatal accidents on all three of the Rockville bridges occurred over time, usually as a result of pedestrians using the railroad as a shortcut and being caught by trains, or when hoboes or others illegally hitching rides on freights fell off and were run over. But this case was out of the ordinary. A Harrisburg-based assistant road foreman of engines, W. B. Steinmetz, 62, was killed on the bridge the night of January 7-8, 1914. Because he was an experienced railroader with 44 years of service, his death was a mystery. He had left for Marysville the evening before and was presumed to have fallen off a train when returning from there.[10]

Signal changes

The signals that directed the flow of railroad traffic to and over the bridge changed around this time, both in type and appearance. An automatic block-signal system, which allowed PRR to retire many manual block stations (towers), first appeared on the railroad in 1889 and was completed westward from Rockville to a point east of Lewistown in 1913. The following year, automatic blocks were in operation between interlocking towers all the way from New York to Pittsburgh and Philadelphia to Washington.[11]

In 1915, PRR first installed Union Switch & Signal Co. position-light signals on the 20-mile Philadelphia-Paoli, Pa., segment of its main line, heavily used by commuter trains. These signals used rows of lights mounted on a black sheet-metal disk to replicate the positions of older semaphore signals, but without moving parts.[12] After the trial installation was successful, the railroad adopted them for systemwide use, and they became a signature symbol of the PRR environment, although few other railroads used them. Later, in the 1930s, with advances in technology, a system of cab signals – in which the lineside signal indications are replicated on a panel inside the locomotive cab, greatly enhancing safety – was installed between Harrisburg and Altoona via Rockville Bridge, and eventually on all of PRR's most heavily used main lines.

During World War I, the federal government seized and directed the operation of most American railroads, including PRR, as a means of keeping military supplies flowing to Europe. The main reason for this was congestion at East Coast ports, and the desire of President Woodrow Wilson to coordinate transportation in support of the war. From December 28, 1917, to March 1, 1920, PRR was run by the U.S. Railroad Administration.[13]

In addition, to protect against sabotage, the Sixth Regiment of the Pennsylvania National Guard encamped at Rockville, beginning in April 1917, and patrolled the area. Soldiers had orders to shoot to kill any canoeists who defied an order to use only the fifth arch from the east side. Another wartime measure to minimize congestion was a reduction in passenger service, including the *Broadway Limited*.[14] It was discontinued from December 1, 1917, to May 25, 1919.[15]

Autos under the arches

Another technological and social development began to nip at the heels of the railroads: the rising acceptance and affordability of the personal automobile. In the early days of motoring and before the coming of a national U.S. route numbering system, boosters of highways gave them distinctive "blaze" markers and marketing names to promote their use and sharpen their identity. In 1916, the William Penn Highway Association was formed in Harrisburg to promote a good cross-state highway that would link important cities that were bypassed by the earlier Lincoln Highway.[16] For much of its length, the William Penn Highway paralleled the PRR main line.

Initially, the route started in Philadelphia, but as finally chosen, it ran from Easton, Pa., on the Delaware River, to Harrisburg, then west to Lewistown (passing under the easternmost arch of Rockville Bridge), Huntingdon, Hollidaysburg, Cresson, Ebensburg, and on to Pittsburgh. Eventually, this became a part of U.S. Route 22. PRR could not have viewed the William Penn Highway as much of a threat, because in those days, motoring from one end of the state to the other took two days, with an overnight stay required somewhere between Lewistown and Ebensburg. Passenger trains made the trip in a mere seven hours.

Another emerging highway was the north-south Susquehanna Trail, which linked Washington with Baltimore, Harrisburg, Williamsport, Buffalo, and Niagara Falls.[17] In so doing, much of it paralleled PRR's Northern Central Railway, passing beneath the westernmost arch of Rockville Bridge at Marysville. Eventually, the Trail became part of the combined U.S. Routes 11 and 15.

As PRR shook off the effects of wartime, federal control, and subsequent labor turmoil, it began to consolidate facilities in the general course of doing business. Unable to expand, Marysville Yard was downgraded in 1921 (but not closed) and most of its business was transferred to Enola.[18] PRR also finally began to rebuild its passenger service in the mid-1920s, adding new trains and services. Along with this direction came a new emphasis on advertising.

PRR and the Osborne Co. calendars

In 1924, William Seely, president of the Osborne Co., a large specialty advertising firm in Newark, N.J., called on an official of the Pennsylvania Railroad in an effort to persuade PRR to begin issuing advertising wall calendars. Given away by businesses of all sizes and types, wall calendars served as a constant visual reminder, keeping a company's name in front of customers year-round in a pre-television age. Small customers such as gas stations and feed mills simply selected from stock styles and illustrations, but prestigious clients commissioned custom artwork. Wall calendars were often large, as big as 28 by 45 inches, because they were intended to be a truly decorative element that could brighten up a dull office or living-room wall.

Harold M. Brett, the Boston artist who painted Speed and Security.
Collection of Ken Murry

below, ***Thrusting Rockville Bridge into prominence again was PRR's Speed and Security oil painting, which appeared on PRR's 1925 and 1926 advertising wall calendars. The canvas measured 56 inches by 93 inches.***
Collection of Ken Murry

The Speed and Security image was used, modified, and reused. top, ***Train brochures for the*** **Manhattan Limited** ***and the*** **Broadway Limited,** ***both New York-Chicago trains, and an east-west timetable;*** above, ***PRR letterhead;*** left, ***a 1929 Christmas card sent by a PRR vice president of traffic, Julien L. Eysmans.***
All, collection of author

PRR had issued calendars sporadically before, but hadn't done so for a number of years. Osborne was selling the idea of establishing a long-running series with a memorable piece of artwork that would occupy a continuing place in the customer's life and mind. In 1922, Osborne had already begun working with the New York Central Railroad on wall calendars that used William Harnden Foster paintings to promote the carrier's *20th Century Limited.* But Seely viewed the PRR as an even bigger prize, if he could land the account. He was successful in doing so, and because the two railroads were rivals, his firm stopped handling the Central account to take up the PRR commission.

Over the years, Osborne had commissioned Harold M. Brett, a New England artist, for various paintings for calendar and other uses. Brett was a friend of another Osborne executive, George C. Hirst.[19] So it was not surprising when Brett was hired to prepare a custom painting featuring a Pennsylvania Railroad train for the railroad's inaugural calendar for 1925. The setting chosen was the east end of Rockville Bridge, with the eastbound *Broadway Limited* crossing it behind K4s-class steam locomotive No. 5357. PRR, or Osborne, titled the painting Speed and Security.

The exact number of calendars produced that year is unknown, but subsequent production quantities ran as high as 300,000. This much is certain, however: the calendar was so popular that PRR exhausted its supply of them and received many more requests than it could fulfill. The company issued calendars that year in two formats. The larger of the two showed all 12 months at once, while the smaller version contained a date pad, with each month to be torn off as it was used. Because of the popularity of this calendar, one of these styles may have been the initial edition, and the other quickly printed as a second edition to meet demand.

In a decision never repeated over the calendar program's next 33 years, PRR chose to re-use the popular scene for its 1926 calendar. The 1926 edition used the tear-off date-pad format and promoted a different train, the *New Yorker* (a New York-St. Louis run). After the initial splash of publicity with the *Broadway*, PRR mated the painting image with the names of numerous other passenger trains in its fleet.[20]

Almost instantly, the Speed and Security painting came to represent the Pennsylvania Railroad itself. It combined a PRR landmark that spoke of stability with a premiere luxury train, pulled by the locomotive type that was rapidly becoming PRR's standard passenger engine—and did so in a recognizable, scenic, real-life setting. Brett's scene used the same formula that New Jersey painter Grif Teller later favored and used so successfully on many PRR calendar paintings: the low-angle head-on perspective with a single fast-approaching train that implied substance, speed, and safety. For 1927, the Osborne Co. again hired Brett to paint a PRR scene, and after that, Teller took over most of the job of producing subsequent paintings for PRR calendars. The series went on for a total of 34 years, and Rockville Bridge was again featured on the 1950 PRR calendar.

During and after the 1925-26 run of the Speed and Security calendar, PRR used the image for a wide array of secondary purposes. Some of these were:

- Wallet calendars.
- Postcards promoting the *Broadway Limited, Liberty Limited,* and the *American.*
- Decks of playing cards, handed out in club cars.
- Timetable covers and advertisements.
- National magazine advertising.
- Dining-car menu covers.
- Named-train brochures for such flyers as the *Broadway* and the *Manhattan Limited* (another PRR New York-Chicago train).
- Gold-foil stamping on the cover of the railroad's 1926 corporate history.
- Service pins that were given to veteran employees for 25, 40, or 50 years of continuous employment.

Obviously, PRR got plenty of mileage out of the Speed and Security image. The same scene, redrawn in line-art form, was used as late as the 1940s on the back of an employees' timetable. The painting was so popular that it was used outside the railroad, sometimes with permission and sometimes in closely copied form. One authorized use occurred when the New Jersey-based Campbell Soup Co. reprinted the painting in a national magazine ad touting its tomato soup. The connection between the PRR and the soup company was not coincidental—from time to time, members of the Dorrance family, which controlled the Campbell firm, sat on the PRR board of directors.[21]

Later, a copied version of Speed and Security—with the bridge edited out, and with freight cars substituted for the passenger train—was used on labels pasted on the ends of Locomotive Engineer brand fruit boxes that were used to transport California grapes.[22] ★

A westbound baggage-and-coach train behind one of PRR's 425 K4s-class 4-6-2-type passenger engines rolls past Marysville Yard, en route to Altoona and Pittsburgh about 1937 or 1938. Excavation in the foreground is preparation for the improved State Route 147 (later, U.S. 11-15) highway. Dave Cope, collection of Ted Xaras

Train 75, the westbound* Fort Dearborn *with 11 cars, rolls across the bridge on August 19, 1933, with K4s-class 4-6-2-type engine No. 5379 in charge. The daily train was scheduled out of New York's Pennsylvania Station at 11 a.m., left Harrisburg at 3:32 p.m., and arrived in Chicago the next morning at 7. Otto C. Perry photo, The Denver Public Library, Western History Department

Streamlining comes to the Pennsylvania Railroad in a big way with the arrival in 1942 of Nos. 6110 and 6111, the first of 52 T1-class 4-4-4-4-type 100 mph steam engines. Collection of Ted Xaras

8. Streamliners, flood, and war

The year 1929 started off as a bright one for the Pennsylvania Railroad. After years of planning, the company on July 7 inaugurated a novel 24-hour transcontinental coordinated rail-air service that passed daily over Rockville Bridge. Passengers bound for the West Coast boarded sleepers of the *Airway Limited* at New York's Penn Station late in the afternoon and rode west through Harrisburg and Pittsburgh, arriving at Columbus, Ohio, the next morning. There, at a trackside airport, they boarded a Ford Tri-Motor plane operated by Transcontinental Air Transport, which took most of the day to fly to Kansas. There, passengers got aboard another train, this time on the Santa Fe Railway, for another overnight sleeper run to New Mexico. In the morning, they caught another TAT plane to complete their journey to Los Angeles. Traveling the whole way by rail took four days. The idea was a creative one for the tradition-bound railroad industry, and other railroads assembled similar rail-air combinations. Although it didn't attract many passengers, it generated publicity. When night flying became safe and practical, the hop-skip-and-jump arrangement quietly faded out of the timetable.

In the Great Depression, all railroads lost money, and many went bankrupt, although not the Pennsylvania. From 1926 to 1932, PRR lost more than half of its freight business, dropping from 244 million tons of freight to 104 million tons. The passenger loss was even more drastic. In 1926, the railroad carried 137 million riders, but in 1932 the figure shrank to 61 million.[1]

Still, even if people didn't ride trains as much, they came out to see special events that occurred on the rails. A visiting British passenger train, the *Royal Scot*, was opened for public exhibition in Philadelphia, then set out on a tour, rolling west over PRR's main line. After taking part in a radio broadcast with PRR's *Broadway Limited* near Harrisburg, on May 20, 1933, it continued west via Rockville Bridge, watched by thousands of trackside spectators. The next year, an even more exotic train passed through Harrisburg. America's first streamlined train was the Union Pacific Railroad's three-car aluminum-bodied M-10000, which was powered by an internal-combustion engine. After an East Coast tour, it left Philadelphia for the Midwest, passing through Harrisburg and over Rockville Bridge on February 18, 1934. Again, people crowded along the tracks, including at Rockville, for a glimpse of the yellow-and-brown speedster, a stark contrast to PRR's conservative tuscan red passenger coaches. As it turned out, streamlined trains were just what the industry needed to bring riders back to the rails, injecting color, excitement, and a touch of luxury into even coach-class travel, and turning travel into an experience again.

In the early spring of 1936, more of both streamlining and internal-combustion came to PRR. The railroad was reluctant to jump into lightweight passenger cars, but did take a K4s-class steam passenger locomotive, No. 3768, into the shops at Altoona and dressed it up with a streamlined shroud

The first streamliner to cross Rockville Bridge was also the first fully streamlined train in America – the Union Pacific Railroad's three-car M-10000. Here it is leaving the east end of Rockville Bridge after departing Harrisburg a few minutes earlier. February 18, 1934.
Collection of Rodney DiPaolo

crafted by industrial designer Raymond Loewy. The shiny engine posed for photos on and near Rockville Bridge, but when it went into service, it ran mostly out of Chicago, with some assignments between Harrisburg and Philadelphia.

At almost the same time, a pair of boxy-looking diesel locomotives working in tandem ran a dozen trial trips during March on regular PRR passenger trains between various combinations of Philadelphia or Paoli, Pa., and Pittsburgh or Chicago, all via Rockville Bridge. Handling eight to 14 cars, locomotives Nos. 511 and 512 came in on schedule time on nine of the 12 trips, turning in a performance of 5,936 miles.[2] They were 1,800-horsepower demonstrator units built by the Electro-Motive Corp., a firm that later became a division of General Motors Corp. By the 1950s, the offspring of these two locomotives had become so commercially successful that they drove steam engines from the rails. But in 1936, PRR viewed diesels as a novelty, and the demonstration trips as a favor to GM, nothing more.[3]

The March 1936 flood on the Susquehanna exceeded even that of 1889. top, ***Water fills Arches 48 and 47 at Marysville.*** above, ***While Front Street in Rockville is submerged, the old canal bed becomes evident.*** right, ***A crane is stationed in the middle of the bridge to break up debris.*** Rockville, Hagley Museum; others, Frank Rupley Daum

The 1936 flood

For the Pennsylvania Railroad, the floods of March 17-20, 1936, were worse than anything that had come before, even the Johnstown Flood of 1889. Hundreds of miles of track were knocked out of service by washouts, bridges that were damaged or carried away, or damage done by floating debris. Flooding in the Ohio, Susquehanna, Potomac, and Delaware watersheds shut down much of the railroad. Water covered some tracks to a depth of as much as 12 feet, and washed out embankments as deep as 130 feet. In all, the railroad reported that as much as 8,271 miles of track, about a third of the system total, had been affected by the flood.[4]

At Harrisburg, the Susquehanna reached a level of 30.33 feet, higher than the previous record of 26.8 feet set in 1889. Tracks in the station were submerged, but as water receded, PRR ran a half-hourly shuttle train across the Cumberland Valley Railroad bridge, which "for several days constituted the only means of crossing the Susquehanna River."[5] Marysville and Enola yards were far enough above the flood to escape damage, but PRR took no chances with Rockville Bridge. "On the afternoon of the 18th the rising water came up to the springing line of the arches some 35 feet from the top, but as trees, buildings, and other heavy objects were coming down the stream, it was thought best to place derrick cars on the bridge, equipped with heavy drop weights to break up the large debris which would not readily pass under the bridge," the railroad later wrote in a summary of the effects of the flood. "This prevented damage to the bridge or further backing up of the flood waters over the surrounding country ...."[6] PRR restored service over most of its lines by March 22.

Special passenger services

Another pair of special trains crossed Rockville Bridge in May 1937. The growth of the railroad enthusiast, or railfan, hobby in the 1930s prompted railroads to operate excursion trains. In an exhibition that likely will never be equaled, PRR held a one-day open house at its Altoona Shops on May 16, 1937, running two excursion trains from New York and one from Chicago to meet at the event. Hundreds of railroad enthusiasts rode the trains to Altoona, where they could view many locomotives that had been shined up and parked for their inspection, before boarding the trains for the trip home.

A little more than a year later, on June 15, 1938, PRR introduced its first four streamlined trains, which it dubbed the "Fleet of Modernism." All four operated via the bridge: the *Broadway Limited* and *The General* between New York and Chicago, *"The Spirit of St. Louis"* between New York and St. Louis, and the *Liberty Limited* between Washington, Baltimore, and Chicago.[7]

West side changes

In 1938 and 1939, significant changes took place on the west side of the bridge. The first was the opening of a new state highway between Marysville and Overview, at the north end of Enola. With railroad tracks occupying the land next to the mountain, highway traffic had traditionally taken a roundabout route, crossing the iron bridge at the north end of Enola, following the river's bank under the westernmost arch of Rockville Bridge and continuing into Marysville, then passing through the subway to the town square. Agitation for elimination of the hazardous road had begun four years earlier with public meetings, at which motorists criticized the many kinks in the road.[8]

Excavation for this project must have rivaled that of Kerbaugh's crews almost four decades earlier. Costing $900,000, or about as much as Rockville Bridge had cost, the project involved slicing 135-foot-deep and 100-foot-deep cuts and removing 320,000 tons of rock. Governor George Earle presided at the dedication ceremonies, which were held on July 21, 1938, at the Marysville PRR station, followed by a parade that included hundreds of automobiles.

The next summer, PRR made its own changes at the west end of the bridge when it finally got around to building a direct connection from the bridge to Enola Yard, a project that had been shelved in the final days of construction in 1902. As built, the original west end of the bridge left space to add a track to

above, ***One of two trains carrying railroad fans from New York and Philadelphia to an open house at PRR's Altoona, Pa., locomotive shops crosses the bridge on May 16, 1937. Two engines are assigned to haul the train.*** below, ***Posed for publicity on the west side of the bridge is PRR's first homegrown streamlined steam engine, K4s-class 4-6-2-type passenger engine No. 3768. The railroad's Altoona Shops gave the engine its smooth shroud in early 1936.*** Both photos, Railroad Museum of Pennsylvania (PHMC)

above, ***An H-class 2-8-0-type steam engine is switching in Marysville Yard.*** Frank Rupley Daum

The divergent paths of the original bridge and the 1902 stone-arch bridge are apparent in this 1930s view of Rockville by Harrisburg aviator/aerial photographer Samuel Kuhnert. Collection of Eric Ready

Enola, but the development of longer and heavier freight cars had rendered that curve too sharp. PRR civil engineers redrew the track alignment to include a modification to the bridge's Pier 47, Arch 48, and the abutment's south wing wall. Crews poured a new concrete foundation on the south side of Pier 47 to hold up the bottom end of a vertical support girder. Alterations were made to the sandstone coping and spandrel walls to accommodate a 119-foot-long steel girder fabricated by American Bridge Co. Its south end rested on an extension of the south wing wall, which was fabricated of concrete above the original sandstone. In all, the modifications cost $178,149,[9] and the project was finished shortly after August 30.[10]

World War II

With the entrance of the United States into World War II, railroading changed, too. PRR, which had not hired any substantial number of people since the 1920s, began hiring in earnest in 1941.

Even before the 1941 bombing of Pearl Harbor, the buildup of defense-related freight traffic strained the limits of the track and interlocking system at the Rockville end of the bridge. The sharp 14-degree curve leading northward to the main line to Dauphin, Sunbury, Williamsport, and Buffalo had limited freight train speeds to 5 mph. In 1940, the number of train movements through Rockville averaged 110 a day, but a year later it had grown to 138. The daily number of trains moving between Enola and the Williamsport Division had risen from 17 to 19. In July 1941, Philadelphia Division officials submitted a proposal to spend $109,963 to ease the curve and replace the mechanical interlocking system at Rockville Tower with an electro-pneumatic system. Both measures would reduce congestion, raising the speed of trains to and from the Williamsport Division to 15 mph and dropping the time each train tied up the bridge from 15 minutes to five minutes. The proposal also included construction of a siding so that southbound freights that couldn't immediately use the bridge to get to Enola because of busy east-west traffic could be stored out of the way of other north-south trains.[11]

The easing of the curve on the wye track to a 10-degree arc was the most involved part of the plan. It envisioned moving all

A westbound passenger train behind another PRR K4s-class 4-6-2-type steam engine is about to pass over the Northern Central Railway tracks. The year is 1938, and PRR has not yet built its direct connection from the bridge to Enola Yard. George Hart photo, Railroad Museum of Pennsylvania (PHMC)

of the PRR tracks on the Williamsport Division eastward to occupy the spot held by the Reading Co. (former Schuylkill & Susquehanna) tracks. Reading trains still used the former S&S branch, but traffic was down to one train a week. The Reading agreed to let PRR swing a wider curve, remove the S&S main track, and move the connection between the two railroads northward about 1,600 feet.[12]

With military fervor running high, PRR posted armed security guards on the bridge, according to brothers Donald, Dominic, and Paul Pinci, whose father, Guiseppe Pinci, was among those so employed. The family then lived on North Main Street in Marysville, and Dominic Pinci recalled visiting his father on the job once when he was about 5 years old. "I wandered away from home one evening," he said. "It was summer. I knew he was down there and wanted to go see him. He was in a shanty and every three hours or so would walk out on the bridge with his revolver and meet a guy who walked across from the Dauphin [County] side. He [couldn't leave his post so he] kept me there overnight and put his jacket over me to keep me warm. In the morning, he sent me to Justone's (another Italian family who lived next to the bridge) for breakfast and then I went home. I thought I was going to get disciplined, but all I got was a good lecture."[13]

Soon the rails were carrying troop trains and war materiel, as well as tank-car trains that were moving oil that could no longer safely evade Nazi submarines in the Atlantic shipping lanes. When the FBI arrested eight Nazi saboteurs who had landed in the United States with plans to blow up PRR's Newark, N.J., station, Hell Gate Bridge in New York City, and Horseshoe Curve near Altoona, it heightened the sense of need for protection.[14]

Goodbye Reading, hello diesels

In November 1944, the Reading Co. filed to abandon the west end of its S&S branch, and after 90 years in operation, the track was pulled up and the connection at Rockville was removed.[15]

By the fall of 1945, PRR's first road diesels were running over Rockville Bridge. The railroad had bought a smattering of diesel-electric yard switchers in the 1930s and early 1940s, but was unwilling to entrust its mainline trains to the new form of motive power. Finally, after every other trunk line in the East (Baltimore & Ohio, Erie, New York Central) acquired mainline diesels, PRR gave in, buying a pair of 2,000-horsepower Model E7 passenger diesels from the Electro-Motive Division of General Motors. The company assigned them to the Harrisburg-Detroit segment of the New York-Detroit *Red Arrow* passenger train. They performed so well that PRR in 1947 abandoned its previous commitment to steam research, development, and operation, and began buying all kinds of road diesels for both freight and passenger service. Within 10 years, diesels would completely replace steam locomotives on the Pennsylvania Railroad.[16]

The company had displaced steam before, but did so with all-electric locomotives. As part of its massive project to electrify its system from New York to Washington and Philadelphia to Harrisburg, the Pennsylvania Railroad had completed electrification into both Harrisburg and Enola Yard in early 1938, and had acquired more than 200 electric locomotives. Company officials repeatedly considered extending the overhead-wire catenary system to Pittsburgh to exploit the efficiency and economy offered by electric operation, but the fiscal climate had changed. PRR no longer had the money to spend, and in fact posted its first-ever financial loss in 1946. Postwar inflation had made labor and materials much more costly than they had been during the Depression, when the

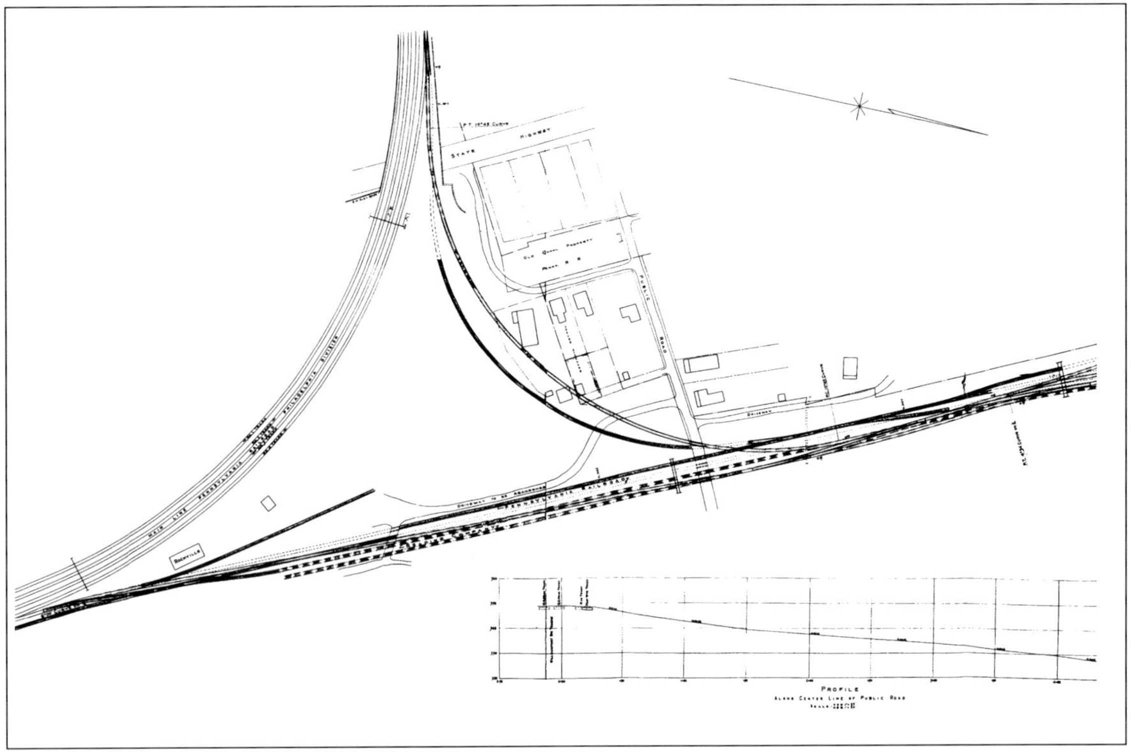

A track diagram shows how PRR realigned its tracks at Rockville to provide a wider curve between the bridge and the main line to the north, speeding wartime freight. The Reading Co. S&S Branch (dashed line) was shortened slightly in the process.
Hagley Museum

The Pennsylvania Railroad's first road diesels were E7-class units No. 5900 and 5901, delivered in August 1945. Quickly and efficiently, these two 2,000-horsepower locomotives changed the way PRR thought about moving its passenger trains. The railroad in 1947 placed orders for a fleet of diesels, and Harrisburg, where they are shown, became their home base. Paul K. Withers collection

PRR featured the bridge on its 1950 calendar painting, Crossroads of Commerce. Artist Grif Teller had taken over from Harold Brett in 1928, painting most of the scenes through the end of the series in 1958. Ken Murry

PRR artist Grif Teller in 1979. Ken Murry

railroad had carried out most of its electrification plans. The result was that poles and wires for electrification never became part of the structure of Rockville Bridge.

Another calendar painting

The transition from steam to diesel locomotives was aptly captured in a painting that artist Grif Teller executed for the 1950 PRR wall calendar. It portrays the junction of tracks at the west end of Rockville Bridge with a westbound diesel passenger train, an eastbound coal train pulled by a steam engine headed for Enola Yard, and a northbound freight train, also pulled by a steam engine, leaving Enola by way of the 1939 connection to cross the bridge and head for the Williamsport Division. Teller had scouted the site on a prewar trip to the area, and used his research photos to give him the right perspective in portraying the scene on canvas. PRR titled the finished product Crossroads of Commerce. But the mixture of steam and diesel power shown on the painting lasted only a few more years. This was one of the last calendar paintings to show steam engines in a prominent role.[17]

Diesels took over the non-electrified parts of PRR in stages. This was a necessity, not a choice, because with 11,000 route-miles of track, the railroad was so large that it was impossible to replace the 4,460 steamers that PRR operated in 1946 all at once.[18]

The Middle Division, Williamsport Division (later, Susquehanna Division, and still later, Northern Region), and Enola Yard were among the places where steam lasted longer than elsewhere, all in regular freight service, being assigned to Enola-Altoona, Harrisburg-Altoona, and Enola-Williamsport assignments. Steam also pulled transfer runs between Harrisburg Yard and Enola Yard, and between Enola Yard and the Reading Co.'s Rutherford Yard in Harrisburg. As a result, Rockville Bridge was the site of one of the last great stands of mainline steam power, not just on the Pennsylvania Railroad, but also in the United States.

Steam engines in regular mainline freight service operated over Rockville Bridge for the last time in the fall of 1957. A Baltimore-to-Northumberland, Pa., rail enthusiasts' excursion that ran on October 20 that year was one of the last, if not the last, steam-powered PRR passenger trains to cross the structure.

Both steam and diesel engines were using a new form of communications that was a predecessor to direct radio transmissions. In 1946, PRR installed an inductive train-phone system on some mainline divisions, including the Middle Division and part of the Philadelphia Division, including the tracks across Rockville Bridge. With this system, which relied on the telephone wires beside the right-of-way as a conductor, engines, cabin cars (cabooses) and towers were equipped with speakers and receivers. On the roofs of engines, tenders, and cabooses, the antennas appeared as handrail-like piping.[19]

An Appalachian Trail reroute

Over the years, the bridge has been used by many unauthorized pedestrians, but the Pennsylvania Railroad opposed a movement that would have made that a regular occurrence. In the 1920s, hikers and naturalists formed a group to create a 2,100-mile-long Maine-to-Georgia hiking path that became the Appalachian Trail. In Pennsylvania, its initial route followed the spine of Blue Mountain, and some early maps informally represented the trail as crossing the Susquehanna on Rockville Bridge. In 1948, a hiker named Earl Shaffer of York Springs, Pa., became the first person to hike the entire trail in one season (he took about four months to make the journey) – an undertaking known as through-hiking and now accomplished by many people each year. In the process, Shaffer became something of a guru on the subject of the Appalachian Trail. With PRR opposing any proposal to formalize the trail's route

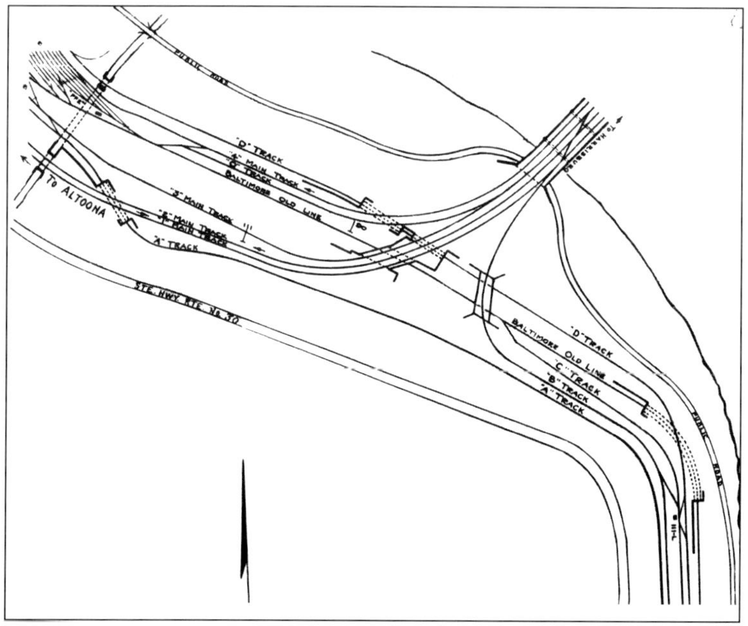

A portion of PRR Philadelphia Division Plan 3796-A, dated June 7, 1946, shows the arrangement of tracks at the west end of Rockville Bridge. Collection of author

PRR commissioned aerial photographer Tom Hollyman to document its main lines, yards, and industrial sites in the 1950s, flying in a helicopter. His view of Rockville Bridge shows two diesel-powered trains about to pass. Pennsylvania State Archives MG-286

left, ***The early 1950s brought a mixture of steam and diesel power to Rockville Bridge as PRR sought to replace its aging steam locomotives. Left, M1b-class 4-8-2-type steam engine No. 6736 begins to lead a train with many stock cars off the bridge and west toward Altoona. Waiting on the track from Enola is an Alco road switcher, ready to cross and turn north toward Williamsport, Pa.*** Don Wood

left, ***In a view from Rockville Tower, an M1b-class 4-8-2-type steam engine rolls an eastbound freight across the bridge.*** Collection of Martin S. Zak

on Rockville Bridge, hikers faced an approximately 10- to 15-mile detour south through Harrisburg and across the Walnut Street Bridge. Shaffer worked with other hikers to draw up plans for a 69-mile relocation of the trail. The rerouting involved moving the trail north through Duncannon to cross the river at Clarks Ferry Bridge, and the new alignment was formally put into use on March 19, 1955.[20]

With the growing postwar use of automobiles and trucks, PRR gradually began to eliminate trains, so that it no longer needed the "Broad Way" of four tracks on main lines. Many were cut back to three or even two tracks, while double-track lines were reduced to a single track. From a systemwide figure of about 5,000 trains a day in the 1930s, the railroad had cut back to 2,800 trains a day by the mid-1950s.[21] By 1956, PRR had rearranged the track and switches on the approaches to Rockville Bridge to eliminate the No. 3 (eastward freight) track in front of Rockville Tower and one track leading into Marysville Yard, which had been downgraded into a storage facility since the 1939 connection to Enola Yard was opened. Still, the bridge carried four tracks and it remained pre-eminent as the largest of PRR's 10,107 bridges.[22]

Another factor that affected the flow of trains through the area was a decision by PRR management to modernize and enlarge the Conway freight classification yard northwest of Pittsburgh at a cost of $35 million. When the final segment of Conway Yard opened on August 4, 1957, that yard replaced Enola as the freight hub of the Pennsylvania Railroad.[23] ★

below, ***PRR M1b-class 4-8-2-type steam engine No. 6736 pulls away from a stop signal and moves its trainload of eastbound coal hoppers onto the bridge and toward Harrisburg.*** upper right, ***New GP9 diesels arrive in Enola Yard with a train containing Pacific Fruit Express refrigerator cars.*** lower right, ***During 1956, a motive power shortage caused PRR to lease several Reading T1-class 4-8-4-type steam engines, including No. 2119, seen here leaving Enola to enter the bridge.*** Below and upper right, Don Wood; lower right, John D. Hahn, Jr.

above, ***PRR L1s-class 2-8-2-type steam engine No. 559 brings an excursion train of railroad fans across the bridge and into Enola Yard, on October 20, 1957. Originating in Baltimore, the train ran behind this engine from Enola to Northumberland, Pa, and back to Enola. Within two months, PRR would convert the last of its steam operations to diesel-electric power.*** John D. Hahn, Jr.

right, ***An M1-class 4-8-2-type steam engine moves coal eastward across the bridge in the early 1950s.*** John Krause, collection of Ted Xaras

left, ***L1s-class 2-8-2-type engine No. 305 brings a freight into Enola that includes a P5a-class electric locomotive, No. 4767. It was not unusual for steam or diesel trains to ferry electric units between Harrisburg and Enola Yards as a convenience when power was needed for trains or units needed to undergo routine inspections. Both yards had electrified trackage, but the only electrified connection between them was by way of the Cumberland Valley Railroad bridge in Harrisburg.*** Collection of Ted Xaras.

below, ***A westbound empty hopper train leaves Marysville Yard behind two Baldwin "Sharknose" freight diesels.*** John D. Hahn, Jr., collection of Walt Ready

The crisp winter air emphasizes details of the plume of smoke rising from the stack of an I1s-class 2-10-0 steam engine, moving freight westward toward Enola Yard. December 1955. Don Wood

Freight train* S-80 *out of Enola Yard is crossing over the east-west Pennsylvania Railroad main line to turn north toward Sunbury, Pa., on March 16, 1963. The locomotives are General Electric U25B 2,500-horsepower unit No. 2521 and Alco RS-27 2,400-horsepower unit No. 2400. Before PRR acquired radio channels and equipment, it used an early form of electronic communication known as trainphone. What appear to be handrails atop the diesels are the trainphone antennas.
Bob Malinoski, collection of Martin S. Zak

9. Competition and merger

The final switchover to diesel-electric power brought to an end 108 years of steam-locomotive operation on the stone Rockville Bridge and its two predecessors. A few very large classes of steam engines had posed an operational wrinkle on the curve at the east end of the bridge. Their extra-long boilers swung wide when the engines were rounding that curve, posing a sideswiped threat, and PRR had to issue a rule prohibiting them from running there when trains were on adjacent tracks. Certain early diesel units also were restricted at that location, but eventually the steam engines were retired and the restrictions on diesels were dropped.[1]

PRR was in poor financial condition in the 1950s, having spent hundreds of millions to upgrade its passenger and freight car fleets and buy new diesels. At the same time, inflation was taking a bigger bite out of its revenues, and the company declared a loss for the second and third times in 1960 and 1962.[2] For a host of reasons, American railroads generally were not earning even a minimal return on investment, and especially Eastern railroads with their higher terminal, real estate, and labor-market costs, and relatively shorter hauls. Thus, PRR began to discuss merging with its old rival, the New York Central Railroad, as a way to combine facilities and cut overhead costs. Talks began in 1957 and after breaking off, resumed in the 1960s.

One of the reasons for railroads' poor financial showing was decisions by federal and state governments to dole out generous subsidies for competing highways and airlines. The passage of the Interstate highway act in 1956 laid the groundwork for construction of a 45,000-mile limited-access highway network across the United States. One of these roads almost came across Rockville Bridge.

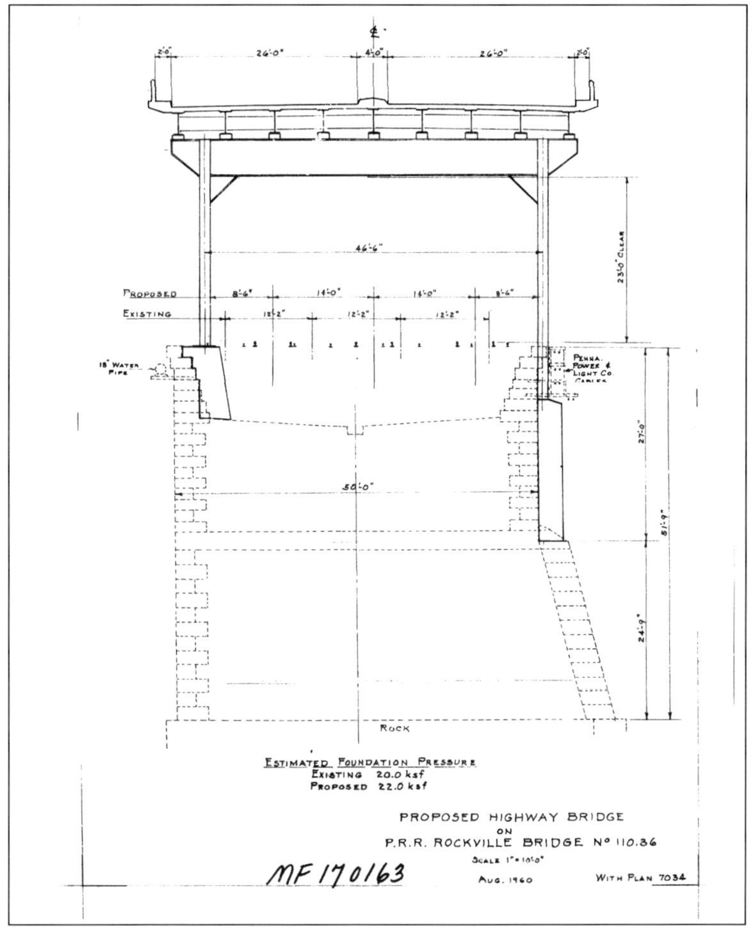

Proposed cross-section for supports to carry Interstate 81 atop Rockville Bridge, dated August 1960.
Collection of Rodney DiPaolo

Pennsylvania already had an east-west toll superhighway, the Pennsylvania Turnpike, opened in 1940, which began to siphon passenger and freight business from PRR soon after World War II was over. It was built with loan guarantees and millions of dollars in outright grants from President Franklin Roosevelt's New Deal program. The success of the Turnpike prompted the growth of more turnpikes after World War II and then stimulated demand for free superhighways.

As the Federal Highway Administration and the Pennsylvania Department of Highways began to consider where to build this network of interstate highways in Pennsylvania, they settled on a north-south route that they named Interstate 81. It would begin at the Canadian border and run through New York, Pennsylvania (where it would connect Scranton, Wilkes-Barre, Harrisburg, Carlisle, Shippensburg, and Chambersburg) Maryland, West Virginia, and Virginia to Tennessee. In the Harrisburg area, it needed to cross the Susquehanna River.

Bridge on a bridge

It's unclear whether the idea originated with highway interests or PRR, but a proposal surfaced to turn Rockville Bridge into a double-decker span – Interstate highway on top, railroad underneath. The four-lane roadway would be carried on a superstructure elevated 23 feet above the rail level. Vertical girders mounted above and partially on the piers would support the highway. At the time, PRR was cash-strapped and was looking at every opportunity to sell air rights over stations, as it did in New York City when it demolished its once-grand Pennsylvania Station in 1963-64.

A westbound passenger train starts to curve off the west end of the bridge in the 1960s. Wearing the next-to-last PRR passenger paint scheme (retaining broad gold striping, but with no PENNSYLVANIA ***lettering), the locomotives are a matched E8A-E7B-E8A set, with unit 5713 in the lead.*** John D. Hahn, Jr.

Word of the stacked bridge idea leaked to a Philadelphia newspaper columnist, whose report was picked up and repeated in the Harrisburg media on February 9, 1961. State highways secretary Park Martin reacted to the report by saying: "So far I haven't heard a word about it. I hope it's true." The supposed benefit of adding a highway bridge to an existing bridge was that "construction time would be speeded up and much of the money needed for a bridge for Route 81 could be diverted to other purposes," *The Evening News* reported. Predictably, PRR denied everything, with Chief Engineer Chester Henry saying "I know nothing about it. It's news to me," and a local PRR spokesman adding that "it would be surprising" if the plan were true.[3] The flutter of news articles quickly faded and when the Pennsylvania Department of Transportation, 1970 successor agency to the Department of Highways, built its own structure for I-81 about a mile downstream of Rockville Bridge. Everybody forgot about the PRR/I-81 proposal.

Nearly 40 years later, a PRR blueprint surfaced in Conrail's Philadelphia headquarters that showed a detailed cross-section of the highway bridge. The plan had been completed by the PRR engineering department and dated August 1960, five months before the newspaper reports came out.

The early 1960s were uneventful for the bridge as the Pennsylvania whittled back its once-extensive east-west passenger service, both in quantity and quality. Even the coming of the Penn Central merger on February 1, 1968, changed very little about the bridge or the traffic rolling over it. The company's tuscan-red passenger diesels were painted black, and the Penn Central monogram, consisting of the interlaced letters P and C, was painted on engines and cars. The sign on Rockville Tower was changed from traditional PRR red and gold to Penn Central's green and white. Despite these and other superficial merger accomplishments, the two former enemies were unable to mesh their economics, operations, computer systems, and corporate cultures. On June 21, 1970, Penn Central declared bankruptcy, the largest corporate failure in American business history up to that time.

above, Westbound PRR SD9 road-switcher unit No. 7610 has swung over to the south track in order to bring its transfer (local yard-to-yard) freight from Harrisburg into Enola Yard on September 8, 1965. right, Powered by two E8 2,250-horsepower passenger diesels, the westbound Juniata*, a New York-Pittsburgh passenger train, crosses over the former Northern Central tracks en route to its first station stop west of Harrisburg at Lewistown, on June 26, 1965.* below, *A long westbound freight crosses the bridge behind GP35 road-switcher unit No. 2287 and three older GP9 units, on November 21, 1964.* All photos this page, Kerry L. Jury

above, *In January 1965, an eastbound PRR freight trails an N5-class cabin car (caboose), one of a fleet of hundreds of N5 cabooses that the railroad built at its Altoona shops. Dating from 1914, the N5 design was the first all-steel caboose type built in America. A rupture in the utility pipe carrying drinking water at Pier 32 has resulted in a frozen waterfall down the face of that pier.*

above, ***Powered by E8 unit No. 4304, a former PRR diesel, an eastbound Penn Central Pittsburgh-New York passenger train passes Rockville Tower on the last day of privately operated intercity passenger service, April 30, 1971. Two freight units are headed west onto the bridge.*** Ken Murry

upper right, ***A derailment of Train** BF-4**, bound from Buffalo to Enola, has brought two railroad cranes into action on June 7, 1969. As a result of Penn Central's poor track maintenance, multi-level automobile-rack cars rocked and swayed when they crossed the connection between the Buffalo main line and the Enola track, and on this occasion some of them derailed. The New York Central car loaded with new Rambler vehicles shown here remained derailed but upright until rerailing efforts began, then it fell over. The cranes are shown picking it up after that happened.*** Kerry L. Jury

lower right, ***What were apparently the first steam locomotives to cross the bridge under steam since PRR closed out mainline steam operations in 1957 were privately owned former Canadian Pacific Railway 4-6-2-type engines 1286 and 1238. They powered a May 18, 1969, circle-trip excursion from Baltimore that ran northward via Perryville, Md., and the Columbia & Port Deposit Branch, to Harrisburg and Rockville Bridge, then returned via Enola Yard and the former Northern Central Railway line through York, Pa.*** Ken Murry

Steam returns, passenger trains diminish

Steam engines pulling excursion trains made a surprising return to Rockville Bridge. In the spring of 1969, a doubleheaded train from Baltimore, pulled by two former Canadian Pacific 4-6-2 engines owned by a private citizen, George Hart, traveled north through Harrisburg, past Rockville Tower and across the bridge before heading south again via Enola. The following year, a leased former Nickel Plate Road 2-8-4 type Berkshire engine, No. 759, ran a pair of Harrisburg-to-Altoona trips on a September weekend.

The final days of privately operated intercity passenger trains came to a close on May 1, 1971, when Amtrak, the Congressionally created National Railroad Passenger Corp., took over service from Penn Central and most other major American freight railroads. Public officials took the action to relieve the financial burden on the freight railroads, which had consistently lost money on passenger business in the wake of hundreds of billions of federal dollars spent on competing highways and airports. The number of passenger trains crossing Rockville Bridge declined from 13 a day to six, and then to four a year later.[4] The final Penn Central-operated passenger train to cross the bridge did so on May 1. It was No. 54, the eastward New York-bound *Pennsylvania Limited,* which had left Chicago on the evening of April 30, before the law took effect. At the east side of the bridge, a ragged, every-other-day remnant of Harrisburg-Buffalo service that passed behind Rockville Tower was discontinued, because Amtrak did not take over that service. Usually consisting of only a locomotive, baggage car, and coach, it was the last remaining passenger train on that route.[5]

Agnes visits Pennsylvania

The formation of Amtrak may have allowed Penn Central to get out from under the economic losses of its passenger service, but it couldn't escape other woes. An unusual sequence of weather patterns in June 1972 brought the company to its knees when unprecedented levels of rainfall destroyed much of its property. Hurricane Agnes began in the Caribbean on June 14 and swept northward, slowing to Tropical Storm status[6] on the way. It went out to sea but turned back, stalling over Central Pennsylvania on June 22, where it discharged four months' worth of rainfall in two days.[7] The Susquehanna rose to its highest recorded level, 32.57 feet (flood stage is 17 feet, and the 1936 flood hit 30.33 feet). The flow of water past Harrisburg was reported as 1 million cubic feet per second.

Tropical Storm Agnes was the most costly natural disaster in the United States until Hurricane Andrew hit in 1992. Damage estimates ran to $3.1 billion, and 117 deaths were reported. Pennsylvania suffered the most, with $2.1 billion in damages and 48 deaths.[8] Penn Central lost 25 bridges, and 23 more were damaged.[9] At Rockville Bridge, the railroad placed loaded coal hopper cars on the tracks, hoping that the weight would keep the bridge intact. The bridge held, but its downstream counterpart, the 28-arch Shock's Mills Bridge on the low-grade freight line, was not so fortunate. It stood during the rainstorm but a few days later, its weakened center section began to sink. On June 27, an engineman spotted the sinking track and alerted railroad officials, who closed the bridge. Some of the piers started to collapse on July 1, and the following day, six arches in the middle of the bridge collapsed into the river.

While Penn Central planned how to attack the problem of rebuilding the bridge, it was forced to reroute the freight traffic – about 50 freights a day – through Harrisburg to Philadelphia via the passenger main line through Lancaster. Many of these operated via Rockville Bridge, temporarily increasing the frequency of trains on that bridge. A rebuilding project at Shock's Mills Bridge started in September when Penn Central dynamited the unstable remaining piers and some adjacent ones, leaving 14 of the original 28 spans, seven on each side. Engineers drew up plans to construct eight new concrete piers as high as 70 feet, which would support nine steel-girder spans.[10] After delays caused by the river washing out construction causeways, the bridge reopened on August 3, 1973, more than a year after it had been closed.[11]

On August 15, 1975, the National Park Service, U.S. Department of Interior, listed the bridge on the National Register of Historic Places. A visual change came to the surroundings on the Rockville side in the late 1970s, when extensive excavation and grading for a new four-lane Route 22-322 highway, which would divert most traffic off the old William Penn Highway route under the easternmost arch. Known locally as the River Relief Route, the new roadway was cut into the hillside behind Rockville Tower. It opened in 1979. ★

The last Buffalo-to-Harrisburg train was Penn Central No. 574, seen here at Rockville Tower on its final run on April 30, 1971, consisting of a baggage car and coach pulled by former New York Central E7 diesel No. 4014. The every-other-day schedule called for eight hours and 10 minutes to make the 312-mile run. Ken Murry

One of the first tasks facing Conrail was the rebuilding of the infrastructure of its bankrupt component lines, including replacement of ties and renewal or replacement of rail. Continuous welded rail was substituted for 39-foot jointed rail in many main lines, and it was transported on special "rail trains" made of specially fitted flat cars with rollers to facilitate loading at the rail mill and unloading at the field site where it was to be installed. Here a Conrail rail train moves eastward across Rockville Bridge. Jim Bradley

10. Conrail, and a costly repair

Penn Central's financial failure and that of several other bankrupt Northeastern lines led to the creation of a government-held railroad to revitalize the system. Congress created the Consolidated Rail Corp. to take over Penn Central, Reading, Erie Lackawanna, Lehigh Valley, Jersey Central, and other lines. As with the Penn Central merger, this change held little difference for Rockville Bridge other than what ran over it. The mixing of locomotive fleets of the predecessor railroads produced strings of color-mismatched diesel units, with, for example, the black of Penn Central, the gray and maroon of Erie Lackawanna, the green and yellow of Reading, and the red of Lehigh Valley.

On August 28, 1977, a collision between two Conrail freights took place on the west end of the bridge, resulting in the derailment of three diesel locomotives and six cars, two of which fell into the river. A southbound Buffalo-to-Enola freight with 103 cars was rolling west across the bridge, crossing over from the north side to the south side to enter the connection to Enola. As it did so, an eastbound 26-car Trailvan (intermodal) train, bound from Detroit to Harrisburg, came around the curve at the west end of the bridge and struck the rear of the other train. The resulting derailment blocked all three tracks.[1] The following day, two tracks were reopened and the third track a few days later.[2]

In the fall of 1979, the American Society of Civil Engineers designated the bridge a National Historic Civil Engineering Landmark. ASCE placed a plaque on the east wing wall during a ceremony held September 27, 1979.

Amtrak continued to operate two daily trains each way over the bridge, the overnight *Broadway Limited* from New York to Chicago and the *National Limited* from New York to St. Louis and Kansas City, Mo. A cutback at Amtrak in 1979 brought the discontinuance of the *National Limited*, leaving only a single daily passenger train on Rockville Bridge, the fewest on the route in more than 125 years. The next spring, the Pennsylvania Department of Transportation inaugurated a state-subsidized daylight daily Amtrak train between Philadelphia and Pittsburgh to fill in the gap in service left by the cut. The *Pennsylvanian* made its pre-inaugural westbound trip on April 26, and regular service started the next day. Eventually, the *Pennsylvanian* was extended eastward to New York and subsequently became the nation's most successful state-sponsored train. It was such a strong financial

On February 2, 1980, an eastbound loaded coal train is powered by SD40 6257, GP40s 3256, 3062, and 3094, and U33C 6867. The units display both Conrail's solid blue paint scheme and predecessor Penn Central's black scheme. Fred Wertz

performer that Amtrak took it back as a national system train, relieving PennDOT of the obligation to subsidize it, and extended its western terminus to Chicago.

Changing tracks, changing trains

Under Conrail management, several changes in signals and the configuration of tracks on the bridge took place. In the 1960s, the main line had been slimmed from four tracks to three, and now in the 1980s, the setup was about to change again. The

upper left, ***Amtrak could not be all things to all train riders, so it offered a program in which states could put up matching funds to add regional trains to the basic Amtrak system. When Amtrak in 1979 discontinued the*** **National Limited*****, one of two long-haul east-west trains through Pennsylvania, the state filled the gap, starting in April 1980, by subsidizing a daylight Philadelphia-Pittsburgh train (later extended to New York), the*** **Pennsylvanian*****. Here the westbound version is crossing Rockville Bridge on November 17, 1991.*** Paul K. Withers

lower left, ***When Congress created Conrail, it also provided for expansion of the Delaware & Hudson Railway, ostensibly to offer competition. D&H, which had been an Albany, N.Y.-to-Canada line with a branch to Binghamton, N.Y., and Wilkes-Barre, Pa., expanded by acquiring a line to Sunbury, Pa., and trackage rights over Conrail to Harrisburg/Enola, Baltimore, and Washington. Carrying American flags, the first southbound D&H train has just crossed Rockville Bridge and is entering Enola Yard on April 2, 1976. Canadian Pacific Railway now operates this service.*** author/collection of Eric Ready

Conrail General Electric U33B unit No. 2916 has just crossed the bridge into Marysville on October 12, 1980, with a westbound "high-and-wide" train, one that requires special handling because of an over-dimension load. Fred Wertz

installation of a centralized Traffic Control System allowed the closing of many lineside block and interlocking towers, including Rockville Tower and Banks Tower (formerly NC, at the west end of Marysville Yard). During the summer of 1986, Conrail double-tracked the main line across the bridge, retaining a third track (the old No. 1 eastward passenger track) on the south side as an extension of the approach track to Enola Yard, known as the "0" or "Aught" (as in zero) track. At the same time, a set of crossover switches between the bridge and Enola Yard, formerly operated from Rockville Tower, was expanded with addition of a second crossover. The installation was designated as a separate interlocking plant and named "HIP," after Wayne Hippensteel, former supervisor of train operations, who had originally been the one responsible for pushing for the installation of crossovers at that point. Conrail also installed a three-track cantilever signal bridge to govern train movements on the bridge, certainly the most striking visual change to the bridge's appearance in many decades. During excavation for the foundation for the signal bridge, a Conrail employee named Don Lauver was killed when he was struck by timber cribbing that fell off a crane sling.

Rockville Tower was closed at 3:01 p.m. on December 9, 1986, but the TCS system was not quite ready for use. In the

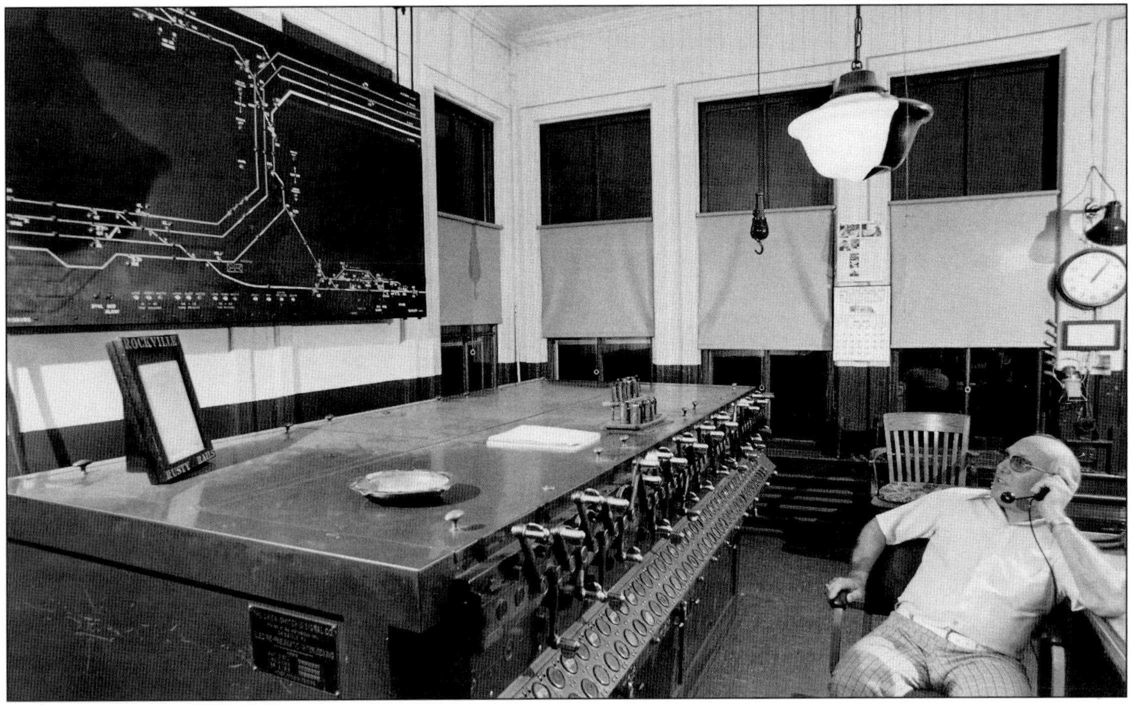

above, ***Third-trick block operator Charlie Longenecker checks the model board (diagram showing position of trains) in Rockville Tower on May 25, 1979.The electro-pneumatic interlocking machine in front of him, installed in 1942, contained the levers that governed switches and signals on both sides of the bridge.*** left, ***Block operator Bill Fry on September 1, 1986, checks the status of a Traffic Control System board that temporarily replaced the electro-pneumatic machine until all dispatching control functions were centralized in an office building a few miles away.*** below, ***Eastbound Conrail Train* TV-4M *pauses at Rockville Tower at 11:48 p.m. on August 31, 1986. With Conrail GP40-2 road-switcher unit No. 3380 leading, it is standing on No. 4 track and bound for Harrisburg Yard.*** All photos this page, Jim Bradley

above, ***Double-stack container freight came to Rockville Bridge in the 1990s.***
Rodney DiPaolo

left, ***After standing guard over the main line for 88 years, Rockville Tower meets its end on September 19, 1986.***
Fred Wertz

intervening two weeks, supervisors equipped with radios had to manually authorize every train to cross Rockville Bridge. The tower stood vacant for almost two years, until it was demolished on September 19 and 20, 1988. Banks Tower near Marysville was demolished on December 9, 1986, the same day that Rockville Tower closed. Another change was made a few years later in conjunction with a 1994-95 program to increase overhead clearances for double-stack intermodal container trains. Dating to soon after the 1902 opening of Rockville Bridge, eastbound freights bound for Enola had to pass through an artificial tunnel built beneath the passenger tracks leading onto the bridge. Railroad cars loaded with double-stacked containers would not fit through the opening, known as Dominic's Tunnel, so Conrail bypassed it by installing a connecting track between the No. 1 (former eastbound passenger main) and the track to Enola near the site of an old switchman's shanty named PF-2. The installation and the site were named MARY, for Marysville.[3]

Besides double-stack trains, other new types of equipment were moving over Rockville Bridge. Before the clearance project was complete, Conrail often moved fully enclosed multi-level automobile carriers between Buffalo and Enola at a time when they could not clear overhead obstructions on the east-west main line. Also in the early 1990s, Conrail opened a small terminal at the former Reading Railroad's Rutherford Yard, just east of Harrisburg, and began operating another kind of intermodal traffic – RoadRailer trains – between there and the Midwest. Unlike trailer-on-flatcar or "piggyback" shipments such as those that began in the 1950s, RoadRailers do away with the flatcar alto-

left, ***As part of a clearance improvement project, Conrail crews install a new interlocking known as "Mary" at the west approach to the bridge in Marysville.*** right, ***The installation, visible with the relay case and signal at upper left, allowed the railroad to abandon a height-restricted tunnel (out of view) that dated from soon after the bridge was opened in 1902.***
Both photos, Rodney DiPaolo

gether. They are composed of highway trailers that either are equipped with a pair of railroad wheels, or, more recently, ride directly on freewheeling trucks (sets of railroad wheels).

Run-through agreements with western railroads to share diesel locomotives on through freights came and went over the years, but they were revived during the Conrail years. On occasion, it was possible for observers in Harrisburg to see such exotic-to-Harrisburg locomotives as the green and yellow of Chicago & North Western, the gray and yellow of Union Pacific, the green and white of Burlington Northern, and the red and silver or blue and yellow of Santa Fe. Conrail also continued the practice of running power through on coal and freight trains from the south. In this traffic, trains came north through the Cumberland Valley from Hagerstown, Md., crossed the Susquehanna on the former Reading Co. bridge at Harrisburg, swung north and then west across Rockville Bridge to enter Enola. This brought engines of Southern Railway, Norfolk & Western, Norfolk Southern, Chessie System, Western Maryland, and CSX Transportation over the bridge.

Amtrak drops the *Broadway Limited*

The *Broadway Limited* New York-Chicago overnight passenger train, successor to the Pennsylvania Railroad's train of the same name, came to its end in September 1995. Under the name *Pennsylvania Special,* the original *Broadway* had made its debut less than three months after the stone Rockville Bridge was opened, so, with two brief interruptions, the train had been running for 93 years. It was the last remaining passenger train over Rockville Bridge to be equipped with a full dining car rather than a snack-bar coach, and the last to carry sleeping cars for first-class passengers. In a budget-cutting move, Amtrak dropped the *Broadway* as a separate run and provided a coach-only connecting train that forced passengers to transfer in Pittsburgh to a Washington-Chicago run, the *Capitol Limited.* Eventually, the railroad reinstated a train on the old *Broadway* schedule but gave it a different name, the *Three Rivers.* A sleeping car was eventually reinstated on that train, too.

Office-car specials, and other excursion trains

Although Conrail was officially a freight railroad, it maintained a fleet of passenger cars and three diesel passenger locomotives for track inspections, special occasions, board of director trips, civic celebrations, and other events. Shippers, government officials, politicians, journalists, and others were the passengers on these trains, which ran under the dispatching symbol "OCS," for office-car special. The dark green diesels and streamlined passenger cars often put in appearances on Rockville Bridge, because it fell between two frequent staging points and two frequent destinations. OCS trains often left from Harrisburg, as the state capital city, or Philadelphia, the railroad's headquarters. Typical destinations for these one-day out-and-back journeys were the locomotive shops at Altoona and the scenic Horseshoe Curve just west of Altoona.

During Conrail's years from 1976 to 1999, several steam-powered excursion trains ran on Rockville Bridge. The most memorable were a series of trips with Strasburg Rail Road equipment, pulled by two former Pennsylvania Railroad steam engines from the Railroad Museum of Pennsylvania collection at Strasburg, Pa. They were Class D16sb 4-4-0 type engine No. 1223, built in 1905, and Class E7s 4-4-2 type engine No. 7002, built in 1902. The routing was Strasburg-Lancaster-Columbia-Enola-Rockville-Harrisburg-Lancaster-Strasburg.

A privately owned Canadian Pacific Class D4g 4-6-0 type engine, No. 972, powered a series of excursions for the Mechanicsburg Museum Association. Among the trips offered was a routing from Mechanicsburg to Harrisburg and then north to turn on the wye at the Rockville end of the bridge. In addition, Blue Mountain & Reading Railroad ran a trip from Read-

Amtrak's final westbound* Broadway Limited *rolls onto the bridge on September 9, 1995. The train and the bridge date from the same year, 1902.
Dan Cupper

ing, Pa., to Harrisburg, with a similar trip to the bridge to turn the train for the return trip to Reading. The engine was former Reading Co. Class T1 4-8-4 type engine No. 2102. BM&R also operated a Reading-to-Horseshoe Curve trip, using two former PRR E8-class streamlined passenger diesels, Nos. 5706 and 5898.

The collapse

Certainly the most infamous incident to occur on Rockville Bridge took place about 8:30 p.m. on August 19, 1997, when a spandrel (side) wall above Pier 19 failed as an 82-car unit coal train was passing over it. The wall and the roadbed above it collapsed into the river at the outside of a curve where the "0" Track connection to Enola diverged from the main line. Four loaded 100-ton hopper cars fell into the river, dumping 400 tons of bituminous coal, along with ballast, masonry blocks from the bridge, railroad ties, and diesel fuel that spilled from a broken fuel line mounted on the side of the bridge. Bound for a Delmarva Power & Light Co. electric generating plant at Indian River, Del., the train under which the bridge fell was symbol *UIR-594*, which had transferred to Conrail from CSX Transportation at Hagerstown, Md. No one was injured in the accident.

Media announcements that "Rockville Bridge has fallen," implying a catastrophic failure of the entire bridge, or even of an arch, turned out to be all hype. Still, the idea of a stone-arch bridge failing was extraordinary. "We're thunderstruck," one federal investigator told the Harrisburg *Patriot-News*.

Underwater inspections in the early 1990s had revealed several piers that exhibited erosion of the underlying concrete and rock foundation. This did not present an immediate problem but, to halt further deterioration, those piers were underpinned with concrete. Inspectors kept an eye on the spandrel walls, but they remained stable.

In 1996, Conrail inspectors noticed movement in some of the stones and implemented daily inspections as a precaution.

top, ***To commemorate the 50th anniversary of the founding of the National Railway Historical Society in Lancaster, Pa., NRHS held its national convention there in 1985 and on June 8, sponsored an excursion from Strasburg, Pa., using equipment from the Strasburg Rail Road and the Railroad Museum of Pennsylvania. Headed by two former Pennsylvania Railroad steam locomotives, Class E7s 4-4-2-type engine No. 7002 and Class D16sb 4-4-0-type engine No. 1223, the train was photographed in approximately the same location as that portrayed in PRR's 1925-26 Speed and Security calendar painting.*** Ken Murry

bottom, ***Running behind Conrail E8 diesel passenger units Nos. 4022 and 4020, an eastbound "OCS," or office-car special for railroad officials, rolls above high water on March 21, 1986.*** Dan Cupper

When the stones didn't move any farther, the inspection frequency was lengthened to weekly, then to monthly. In April 1997, inspectors detected the stones moving a quarter-inch to a half-inch between their inspections and submitted a repair plan to Conrail headquarters in Philadelphia. The plan was approved and work was set to begin on August 22 when inspectors, during that same month, found that overnight, some stones had moved 2-1/2 to three inches. They recommended that the bridge be closed immediately, but with the contractor already scheduled, Conrail management overruled that proposal.

After the collapse, Conrail immediately closed the bridge, but determined that it was stable enough to reopen the No. 2 (north) mainline track just four hours later. Railroad-tie cribbing was built to shore up the roadbed at the point of failure, and the No. 1 (south) mainline track was reopened three days later.

The immediate repair to allow the resumption of freight traffic on the adjacent main track was to construct a timber crib to retain ballast, backfill this wall and construct the track over it. Later, the failed portion of the spandrel was reconstructed with reinforced concrete, using architectural formwork to mimic the original stone construction of the bridge.

During the disruption, some trains detoured to and from Enola

left, ***Tracked-vehicle side-boom cranes move into position to remove the four CSX Transportation 100-ton-capacity hopper cars that fell into the Susquehanna.*** below, ***Timber cribbing is in place and backfilling has begun.***

left, ***Steel rail is twisted like licorice on the "0" or "Aught" track after the spandrel wall failure of August 19, 1997.***
All photos this page, Mark Brinck

An architectural-replication services contractor has placed wooden forms at Pier 19 to allow cement to be cast with a texture matching that of the original sandstone. Rodney DiPaolo

A view of the finished repair patch at Pier 19 on January 28, 2002, shows the finished pattern and the difference in color between weathered sandstone and unweathered concrete. Dan Cupper

via a reversing move at Banks Interlocking west of Marysville. In addition, some trains were detoured north via Lock Haven, Pa., and the Nittany & Bald Eagle Railroad. Conrail reopened the Enola connection by moving the switch several hundred feet west on the bridge while repairs to Pier 19 continued.

To minimize environmental damage, an environmental contractor placed containment and absorbent booms around the derailed cars in the river. Additional booms were placed at City Island in Harrisburg and further downstream at Steelton, after state Department of Environmental Protection officials noticed a fuel oil sheen at both places, both of which were near intake pipes for utility or municipal drinking water suppliers. No fish kills were observed.[4]

The river is so shallow at this point – about three feet in dry weather – that it was possible to drive cranes into the river to remove the hopper cars. In August, two tracked sideboom cranes operated by Hulcher Services lifted the cars off the riverbed and dragged them about a half-mile downstream to a landing on the east shore. This method posed the least environmental impact and allowed United Water Co. to continue using its intake pipe.[5] About a month after the collapse, Conrail began removing the coal and other debris from the river. The railroad originally proposed doing so with a crane positioned on the bridge, but decided against that for safety reasons. Instead, the company hired a contractor, Conti Enterprises, to remove the coal and debris with trackhoes (tracked-vehicle backhoes) and dump trucks.[6] DEP later fined the contractor $9,450 for causing sediment pollution during the cleanup. Originally, Conti had planned to float a barge to the site, but the water level was so low that it excavated a channel in the river. The method of doing so, DEP found, caused the pollution.

Conrail addressed the issue of pier integrity by adding steel-girder bracing on either side of the piers, tied together with steel rods piercing the masonry. The Pennsylvania Railroad had taken this approach with several other stone-arch bridges in the 1960s, but Rockville Bridge had never needed that treatment. In the mid-1960s, a program had been conducted to secure several spandrel walls that had shown signs of movement, mostly along the south side, but it was addressed differently. The work involved excavating behind the spandrel, anchoring in reinforcing bars and pouring a mass of concrete to prevent further outward movement of the spandrel stones. In the 1997 repair, Conrail added channel braces to Piers 6, 17, 18, 19 (upstream side only on Pier 19; downstream side has concrete patch), and 20.

Meanwhile, Conrail considered how to repair the broken spandrel wall, which was caused by the freeze-thaw cycle breaking apart the means by which the sandstone blocks were tied together. The location at the outside of the curve on the Enola connection may have been a contributing factor, but engineers concluded that the freeze-thaw cycle was primarily at fault. That situation was made worse by inadequate drainage inside the bridge, which was built with only two "weepholes," or drain pipes, per arch.

Retrieving the fallen masonry and reassembling it was out of the question, because some blocks were broken. The railroad finally decided to cast a new concrete spandrel wall with facing to match the texture of the surrounding sandstone. Conrail hired a contractor, who took a molding off the east abutment to use as a pattern. In a decision that not everyone liked, a Conrail engineer chose the color of the concrete to match not the dark weathered surface, but the original white or whitish-pink hue of freshly cut stone, on the premise that it would eventually look like the rest of the bridge. Sometime after the patch was in place, Conrail re-extended the "0" Track Enola connection along the south side of the bridge and joined it to the main line at its previous location. In the end, the repairs cost Conrail about $1 million, or the same as the original construction cost.

Norfolk Southern takes control

On June 1, 1999, the bridge became property of Norfolk Southern Corp. when NS bought 58 percent of the company and acquired most of the former Pennsylvania Railroad portion of

Conrail. CSX bought the other 42 percent and in so doing acquired many former New York Central Railroad lines.

Rockville Bridge appeared on a third railroad-issued calendar in 2000 when Norfolk Southern featured it in a cover photograph. The company, which uses photos taken by its employees in its calendars, chose a scene showing the bridge at sunset with wildflowers, shot by Winston D. Frazier, a transportation department clerk based at Kenova, W.Va.

NS in 2001 eliminated the south track on Rockville Bridge by shortening the Enola Yard "0 Track" connection, moving the switch where it joins the main line back to the west end of the bridge, near where it had been originally located in 1939. The reason was to reduce the stresses on the outside of the bridge, minimizing the chances for a recurrence of the 1997 failure. This change was effective November 12, 2001.[7]

Norfolk Southern also strengthened three additional piers in the same way that Conrail had done in 1997, by adding steel-girder bracing to Piers 21, 40, and 41.

After the terrorist attacks of September 11, 2001, Norfolk Southern, like many railroads and institutions, grew more security-conscious. Effective October 27, 2001, NS assigned special agents (police officers) to patrol the east and west approaches to the bridge.

In the fall of 2001, the Altoona Railroaders Memorial Museum conducted its customary autumn Railfest weekend celebration, which included a series of local trips around Horseshoe Curve. To handle the passengers, the museum leased a group of commuter coaches owned by the state of Maryland, pulled by two E8-class former Pennsylvania Railroad streamlined passenger diesels, restored to their 1956 appearance by Philadelphia entrepreneur and PRR historian Bennett Levin. Assembled in Philadelphia, the train crossed Rockville Bridge on its way to and from the event, bringing a touch of history to the structure once more.

To commemorate the 100th anniversary of the opening of the stone-arch bridge, a group of Marysville-area residents organized a Rockville Bridge U.S. Stamp Committee to promote the idea of a commemorative postage stamp to be issued by the U.S. Postal Service. The committee enlisted many citizens as well as public officials to write letters in support of the proposal. While the USPS did not issue such a stamp, the group did succeed in raising public awareness of the significance of the bridge and its 100th anniversary. ★

Two views from July 1999 highlight the mixture of Norfolk Southern and Conrail locomotives that followed NS's June 1, 1999, acquisition of 58 percent of the Conrail trackage, including most lines in Pennsylvania. above, ***Conrail SD70 unit No. 2571 leads an NS Dash 9-40CW unit and a Conrail GP15-1 with a westbound train having a long string of gondolas at the head end. In late March 2002, Norfolk Southern removed the cantilever signal bridge that had stood since 1986, replacing it with lineside mast-mounted signals.*** right, ***Conrail SD60I unit No. 5611, assembled at Conrail's Juniata Locomotive Shop in Altoona, Pa., in 1995 and NS Dash 9-40C unit No. 8820 make a near-mirror image in the water as they head east.*** both photos, Brian Plant

A Rockville Gallery

Collection of Duane Hammaker

Collection of Rodney DiPaolo

Collection of author

Collection of author

Collection of author

Collection of author

Collection of Eric Ready

Speed and Security

The Speed and Security painting image, used for PRR's 1925 and 1926 calendars (*left*, 1926 example) was reused for many purposes. bottom left, ***A paper fingerbowl liner announced the Broadway Limited.*** below, ***Employees who served 25, 40, or 50 years received a service pin that bore a copy of the scene.*** right, ***Campbell Soup Co. used it to emphasize the theme of a parallel of quality in food and transportation. This advertisement appeared in the Saturday Evening Post on January 23, 1926.*** bottom right, ***In California, thousands of miles from the nearest PRR rails, Locomotive Engineer brand fruit labels display the borrowed perspective and lines of Brett's painting.*** Pin and advertisement, collection of author; all others, collection of Ken Murry

above, ***Eastbound and westbound freights, both headed by I1s-class 2-10-0-type steam engines, pass at the west end of Rockville Bridge on August 28, 1955. right, A helper engine is pushing on the rear of the same westbound train.*** Both, Walter Minnich Jr.

below, ***An eastbound passenger train is crossing above the former Northern Central Railway tracks that connect Marysville with Enola Yard, headed by a pair of tuscan-red E7-class diesels (PRR class EP20), with No. 5845 leading, on June 19, 1956.*** Walter Minnich Jr. right, ***For a year and a half in 1956 and early 1957, PRR operated a fast General Motors lightweight experimental passenger train named Aerotrain, whose coaches were modified GM bus bodies, on a daily New York-Pittsburgh round-trip schedule. Passengers' opinions were widely mixed, and its reliability was shaky. PRR sent the train back to GM and declined to place an order for any production copies of it. Here the westbound version leaves the west end of the bridge.*** Don Wood

right, ***After the demise of the Pennsylvania Railroad, some of its first-class passenger cars and office cars were sold to private owners, who restore and operate them for personal, excursion, or charter trips. Here the*** Mountain View, ***one of two 1949 Pullman-Standard observation cars containing two master rooms, a bedroom and a buffet-lounge section built for the*** Broadway Limited, ***brings up the rear of Amtrak's eastbound*** Pennsylvanian ***at Rockville in October 1991.*** Dan Cupper

above, ***Amtrak's eastbound*** **National Limited** ***crosses the bridge in 1975 behind an SDP40F unit and an older E-series booster unit.*** Dan Cupper

above, ***The 48 cars of the Ringling Brothers/Barnum & Bailey Circus "blue unit" train stretch across almost the entire length of the bridge at 11:50 a.m. on June 11, 1990. Moving under the Conrail symbol*** **CIR-101**, ***the train is being hauled by No. 6423, an SD40-2 model, and No. 3275, a GP40-2 model.*** John W. Smith left, ***A Chessie System empty hopper train pulls out of Enola Yard and across the bridge on October 29, 1983. The ruins of the piers of the former bridge appear at right.*** Paul K. Withers below, ***The swollen Susquehanna has risen to near the springing line (the point where the bridge arches meet the piers) as Amtrak's overnight*** **Broadway Limited** ***from Chicago rolls toward Harrisburg on February 13, 1981.*** Fred Wertz

Locomotives on Rockville Bridge can be multicolored and powerful, or plain and small. upper left, ***Conrail SW1200-model diesel switcher No. 9336, a former Penn Central/Pennsylvania Railroad unit, trails six arches' worth of white smoke as it moves eastward across the bridge on April 3, 1980.*** Fred Abendschein upper right, ***Santa Fe red-and-silver and blue-and-yellow units spice up the scene on October 13, 1991, as Train ENBU (Enola-Buffalo) moves out of the yard with 144 cars. The diesel road units are Conrail SD40-2 No. 6436, Santa Fe SD45 No. 5307, Santa Fe FP45 No. 98, and Conrail GP35 No. 2275.*** Fred Wertz

bottom left, ***A special "Farewell to Conrail" charter train made up of 14 privately owned rail passenger cars moves eastward off the Harrisburg-Pittsburgh main line at Rockville, turning northward to enter the Buffalo Line on October 2, 1998. The locomotives leading the way are Amtrak P42DC Genesis unit No. 109 and F40PH unit No. 391.*** Fred Wertz bottom right, ***Three General Electric Dash 9 units, Nos. 8947, 9639, and 9340, are pulling a 105-car Norfolk Southern freight eastward out of Enola Yard and across the 1939 steel girder extension to enter Rockville Bridge on January 28, 2002, headed for the Buffalo Line.*** Dan Cupper

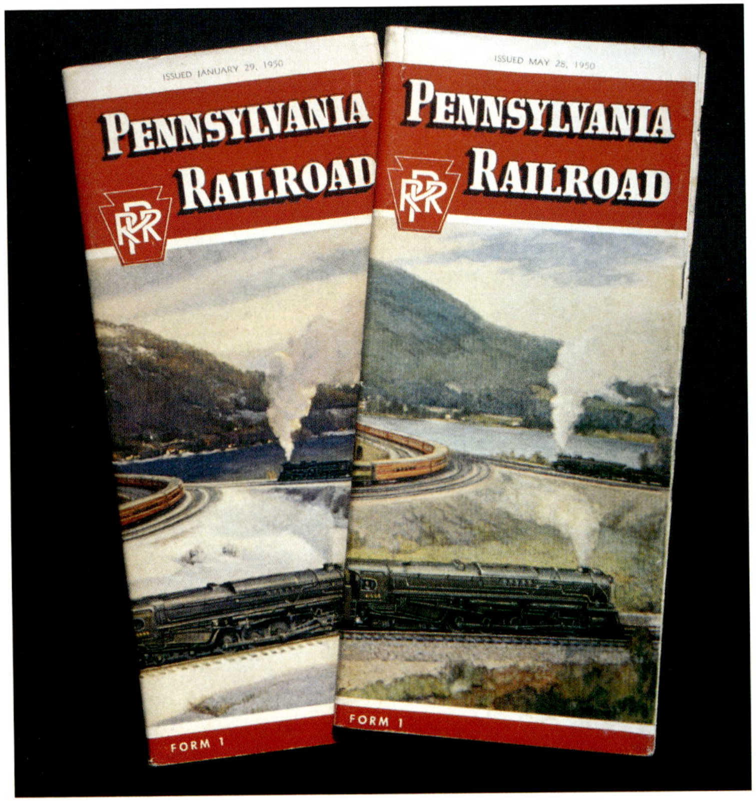

above, ***Form 1 PRR timetables show summer and winter versions of the Crossroads of Commerce 1950 calendar scene.*** collection of author below, ***Appalachian Brewing Co. in Harrisburg uses a stylized illustration of the bridge in its advertising and on its trucks and beer bottle labels. Other mementoes of the bridge include textured postcards, right, and paperweights, lower right.*** Both, collection of Rodney DiPaolo

With prominent municipal (top center), ***bank*** (above) ***and state historical*** (below) ***signs, a visitor to Marysville soon gets the idea that Rockville Bridge must be somewhere nearby.*** Three photos, Dan Cupper

Appendix A – Accidents during construction of stone Rockville Bridge

Four men and two boys, both Italian water boys, died in the construction of the bridge. Fatalities listed in **boldface** type

Date	*Worker's name and nationality/race, if known*	*Contractor*	*Injury*	*Notes*
About July 11, 1900	Lewis Potot, black	Drake & Stratton	Several toes of left foot mashed when run over by a truck.	Hospitalized at city hospital; part of foot amputated.
July 23, 1900	Salvator Metazapo, Italian	Kerbaugh	Fracture of left hand.	Treated at city hospital, released.
August 1, 1900	**Angelo Mucci,** Italian	Kerbaugh	Struck by dinky engine, body cut in half.	Died; buried on August 3 in Catholic Cemetary, Harrisburg; left wife and three children in southern Italy.
October 17, 1900	Harry I. Brady	Kerbaugh	Left toe crushed.	Treated at Harrisburg Hospital; toe amputated.
October 20, 1900	Robert Carr	unknown	Several toes injured.	One toe amputated at Harrisburg Hospital.
April 8, 1901	**G. Pasqualo**	Kerbaugh	Crushed by rock fall.	Lived two minutes after being struck by rock. Buried in Catholic Cemetary, Harrisburg.
May 14, 1901	Amos Miller, 19	Kerbaugh	Fell into river.	Rescued by Howard Keim, 19, of Marysville.
June 11, 1901	Mickele Ballerinni, Italian	Kerbaugh	Knocked down by dinky engine; three broken bones in right foot.	Admitted to Harrisburg Hospital
June 21, 1901	**Recell Ross/Ress Recell**, 13, Italian water boy	Kerbaugh	Walked in front of dynamite blast.	Admitted to Harrisburg Hospital; died at midnight of the same day.
September 11, 1901	**Richard Davis,** 20 (Baltimore)	Drake & Stratton	Struck by draft of narrow-gauge cars about 8 a.m.	Died at site about 10 a.m.
About September 10, 1901	Joseph Vlonjnic	Kerbaugh	Fell; suffered contused back.	Kept at workers' quarters until condition worsened; then transferred to Harrisburg Hospital.
September 28, 1901	Paul Lee, black (Virginia)	Drake & Stratton	Struck on head by wrench while working at cement mixer.	Wrench had fallen 30 feet.
December 19, 1901	Lillen Gumatz/Lellen Grumatz	Kerbaugh	Fell from train being unloaded, bruised hip.	Taken to Harrisburg Hospital on shifting engine with Ulconec.
	Mille Ulconec	Kerbaugh	Fell from train, bruised hip immediately after Gumatz's accident.	Taken to Harrisburg Hospital on shifting engine with Gumatz.
January 18, 1902	**Nicholas Sabastino,** 12, Italian	unknown	Struck by train; body cut in half.	Died instantly.
February 16, 1902	Mike Bi Ioar, 47, Italian	Kerbaugh	Leg broken in two places when he slipped while boarding a workmen's train.	Admitted to a hospital in Harrisburg.
April 16, 1902	**Unknown foreign workman**	unknown	Struck by eastbound passenger train.	Died instantly.
August 30, 1902	Harry Douglass, 32	Kerbaugh	Right leg fractured and suffered other injuries when he fell 40 feet over an embankment at the west approach to the bridge.	Admitted to Harrisburg Hospital.

Names and details are drawn from contemporary newspaper accounts, which did not always match. Harrisburg's three daily papers covered laborers' accidents inconsistently. One, two, or all three papers might report an accident, but they might differ in details or spelling of names. This list is limited to accidents related to the 1900-02 construction of the stone Rockville Bridge. Excluded are accidents and fatalities that occurred in the related excavation and construction, by the same contractors, of the main line at Marysville, the main line at Duncannon, and Enola Yard. Also excluded are accidents occurring to those who were not employed on the bridge.

Endnotes

Chapter 1

1 Democratic Union, 21 July 1847
2 History of Susquehanna and Juniata Valleys, p. 440
3 PRR First Annual Report, p. 24
4 C&C Vol. 1 p.47
5 C&C Vol 1, p. 67
6 Guide to the Pennsylvania Railroad, 1855
7 PRR Second Annual Report, p. 11
8 Haupt on Bridge Construction, p. 170
9 Democratic Union, 31 March 1847
10 Haupt on Bridge Construction, p. 170
11 Description and History of Important Bridges and Stations, 1945
12 Haupt on Bridge Construction, pp. 169, 171
13 Democratic Union, 25 July 1849
14 History of Susquehanna and Juniata Valleys, p. 441
15 PRR Schedule No. 10, Eastern Division, May 16, 1853
16 PRR Eighth Annual Report, for year 1854, p. 42
17 PRR Ninth Annual Report, for the year 1855, p. 35
18 PRR Ninth Annual Report, for the year 1855, p. 33
19 PRR Sixth Annual Report, for the year 1853, p. 48
20 PRR 11th Annual Report, for the year 1857, pp. 46-47
21 PRR 11th Annual Report, for the year 1857, p. 47
22 PRR 13th Annual Report, for the year 1859, p. 39
23 Sources disagree on the number of spans, with some stating the number as 14 or 18; NCR Annual Reports consistently refer to 19 spans or allude to repairs on a Span 19.
24 Story of the Northern Central Railway, pp. 39, 182
25 PRR 12th Annual Report, for 1858, p. 40
26 NCR 10th Annual Report, for 1864, p. 12

Chapter 2

1 From Walker's entry, "Jackson's Capture of Harper's Ferry," in Battles and Leaders of the Civil War
2 Patriot and Union, June 29, 1863
3 Patriot and Union, June 18, 1863
4 Hare, History of the Reading, p. 142
5 Patriot and Union, June 29, 1863
6 "Rebels in Pennsylvania," by Uzal Ent, in Civil War Times, August 1998
7 An organized Union Army force of soldiers who had been disabled by weapons, disease, or accidents, some of whom were still capable of bearing arms or performing garrison duty.
8 Official Records of the War of the Rebellion, Series 1, Vol. 27, Part 2, Page 230
9 Patriot and Union, July 7, 1863
10 Official Records of the War of the Rebellion, Series 1, Vol. 27, Part 2, Page 230
11 Telegraph, July 18, 1868
12 Telegraph, July 29, 1868
13 Telegraph, July 20, 1868
14 Patriot, July 29, 1868
15 Patriot, August 17, 1868
16 Telegraph, August 17, 1868. The Patriot reported that the train consisted of 49 cars; in his accounting, the Telegraph writer may have omitted the baggage cars, because his count for passengers – about 100 per car – matches the count for the number of coaches he reported.
17 PRR 23rd Annual Report, for year 1869, p. 39
18 PRR 21st Annual Report, for year 1867, p.48
19 C&C Vol, 1, p. 67
20 Sipes, p. 26
21 NCR 18th Annual Report, for year 1872, p. 27
22 Telegraph, April 4, 1872
23 Baldwin Locomotives Magazine, October 1934, p. 19
24 C&C, Vol. 1, p. 80
25 PRR 30th Annual Report, for year 1876, p. 30
26 Sipes, pp. 31-32

Chapter 3

1 Patriot, July 16, 1877
2 PRR 31st Annual Report, for 1877, p. 166
3 C&C Description and History of Important Bridges and Stations 1945
4 Patriot, June 2, 1877
5 Patriot, June 15, 1877
6 PRR 31st Annual Report, for 1877, p. 166
7 Annual Report, Adjutant General of Pennsylvania for 1877, pp. 90-91
8 Patriot, August 9, 1877
9 Telegraph, August 12, 1901
10 Patriot, September 5, 1877
11 Patriot, October 20, 1877
12 An article published in Baldwin Locomotives Magazine, October 1934, reported that 21 spans measured 156 feet eight inches and two spans were 150 feet 1-1/2 inches.
13 Patriot, September 11, 1877
14 Letter dated June 6, 1942, governing change of track at Rockville, Hagley
15 Telegraph, April 24, 1882
16 Telegraph, May 4, 1882
17 Telegraph, May 9, 1882
18 C&C Vol. 1, p. 88
19 NCR 28th Annual Report, for 1882, p. 24
20 NCR 30th Annual Report, for 1884, p. 54
21 Unsigned note "Cost of Renewals Rockville Bridge" dated August 15, 1883, Hagley
22 Letter from M. Riebenack to W.H. Brown dated October 15, 1883, Hagley

Chapter 4

1 Harper's Weekly, January 20, 1877
2 Tyrrell, p. 95
3 PRR 31st Annual Report, for year 1877, p. 167
4 Pennsylvania Historic Railroad Bridge Project, Historic American Engineering Record, National Park Service 1999-2001, Vol. 1, HAER No. PA-530
5 Patriot, August 1, 1900
6 June 3, 1899, letter, Hutchinson to Brown, Hagley
7 Patriot, November 15, 1899
8 Star-Independent, November 2, 1899
9 Star-Independent, March 9, 1900
10 PRR 53rd Annual Report, for year 1899, p. 18
11 Patriot, March 6, 1900
12 Patriot, March 8, 1900
13 Telegraph, March 23, 1900
14 Star-Independent, March 9, 1900
15 Star-Independent, April 21, 1900
16 Telegraph, March 14, 1900
17 On the west, the curve over Arches 45-48 was 5 degrees; on the east, over Arches 1-4, it was slightly tighter, at 6 degrees.
18 The Railroad Gazette, October 25, 1901, p. 732
19 The Railroad Gazette, October 25, 1901, p. 732
20 Star-Independent, April 13, 1900
21 Star-Independent, April 13, 1900
22 Telegraph, March 20, 1900
23 Patriot, April 16, 1900
24 Telegraph, March 31, 1900. The firm also did masonry bridge work for PRR; in May 1900 it received a contract to widen a stone-arch bridge on the main line near Port Royal, Pa. Telegraph, May 12, 1900.
25 Patriot, March 21, 1900
26 Patriot, April 6, 1900
27 Telegraph, March 31, 1900
28 Star-Independent, April 9, 1900
29 Patriot, April 11, 1900
30 Star-Independent, April 4, 1900
31 Star-Independent, April 13, 1900
32 Star-Independent, April 13, 1900; May 28, 1900
33 Telegraph, April 28, 1900
34 PRR Drawing 7034, the main blueprint covering arches and east end details, is dated April 6, 1900.
35 Patriot, April 13, 1900
36 Telegraph, April 28, 1900
37 Patriot, May 2, 1900
38 Star-Independent, May 7, 1900
39 Star-Independent, May 7,1900. The paper incorrectly reported Nauman's name as Norman; thus, the hotel's nickname was likely misreported as well.
40 The Railroad Gazette, October 25, 1901, p. 732
41 Star-Independent, May 28, 1900

[42] Railroad Gazette, October 25, 1901, pp. 732-733
[43] Patriot, June 9, 1900
[44] Star-Independent, June 29, 1900, and Telegraph, July 23, 1900
[45] Telegraph, July 23, 1900
[46] Telegraph, July 31, 1900
[47] Telegraph, August 16, 1900
[48] Patriot, June 9, 1900
[49] Telegraph, August 2 and 3, 1900
[50] Telegraph, August 22, 1900
[51] Telegraph, September 12, 1900
[52] Telegraph, September 5, 1900
[53] Telegraph, September 5, 1900
[54] Star-Independent, September 26, 1900
[55] Star-Independent, October 2, 1900
[56] Star-Independent, October 13, 1900
[57] Star-Independent, November 24, 1900
[58] Telegraph, January 22, 1901

Quarries - page 33

[1] Star-Independent, May 15, 1900, and Telegraph, September 8, 1900. The grade was 52 feet in 100, or a 52 percent grade. The Patriot of May 17, 1900, gives the length of the inclined plane at 12,000 feet, more than two miles, and is likely in error.
[2] A notation on PRR engineering drawing 7034 identifies a source for the stone as being Grampian, Pa., which lies about four miles south of Greenville, the tiny village that is closest to Bloom's Run Quarry. Whether the notation is intended to identify that locale as the exclusive provider of sandstone for the bridge is unknown; numerous other sources identify other quarries as producing stone for the bridge.
[3] Quarries of Curwensville: The People, The Legacy, p. 68
[4] The Railroad Gazette, October 25, 1901, p. 732

Brown - page 35

[1] Journal of the Lancaster County Historical Society, Vol. 44, No. 4 (1940), pp. 58-59
[2] History of the Pennsylvania Railroad, Wilson, pp. 39-40
[3] Lancaster Inquirer, March 3, 1906

The Contractors - page 37

[1] Bowers v. Kerbaugh-Empire Co., 271 U.S. 170 (1926)
[2] Star-Independent, April 26, 1900

RJ Tower - page 45

[1] When PRR later dropped telegraphic symbols for many towers, RJ was renamed Rockville.
[2] Telegraph, September 9, 1898
[3] Telegraph, October 5, 1901
[4] 1952 PRR Philadelphia Interlocking book

Chapter 5

[1] Telegraph, January 26, 1901
[2] Telegraph, May 15, 1901
[3] Letter to Chief Engineer W.H. Brown from Superintendent of Telegraph A. Heale, dated March 5, 1901, Hagley Museum.
[4] Telegraph, March 28, 1901
[5] Patriot, April 2, 1901
[6] Telegraph, April 5, 1901
[7] Telegraph, April 3, 1901
[8] Star-Independent, April 6 and April 8, 1901
[9] Star-Independent, June 3, 1900
[10] The line would get even closer in 1914, when the company extended its track all the way to Rockville.
[11] Telegraph, March 17, 1900
[12] Patriot, April 3, 1901
[13] Telegraph, June 22, 1901
[14] Telegraph, April 11, 1901
[15] Telegraph, October 12, 1901
[16] Telegraph, October 12, 1901
[17] Telegraph, April 11, 1901
[18] Telegraph, May 15, 1901
[19] Telegraph, May 18, 1901
[20] Star-Independent, July 14, 1902
[21] Telegraph, May 15, 1901
[22] Telegraph, May 18, 1901
[23] Telegraph, September 27, 1901
[24] Telegraph, October 10, 1901
[25] Telegraph, June 19, 1901
[26] Telegraph, July 9, 1901, August 12, 1901
[27] Star-Independent, September 9, 1901. The Telegraph reported that the same incident occurred on Sunday, September 8.
[28] Patriot, September 17, 1901
[29] Patriot, September 18, 1901
[30] Correspondence, Vulcanite Paving Co. to PRR Chief Engineer W.H. Brown, dated September 24, 25, 26, 28, 1901, Hagley
[31] Telegraph, October 10, 1901
[32] Telegraph, September 28, 1901
[33] Telegraph, October 26, 1901
[34] Telegraph, December 3, 1901
[35] Telegraph, January 7, 31, 1902
[36] Telegraph, December 16, 1901
[37] Telegraph, May 23, 1902
[38] Star-Independent, December 20, 1901
[39] Star-Independent, December 27, 1901
[40] Telegraph, February 17, 1902
[41] Star-Independent, February 22, 1902
[42] Telegraph, March 13, 1902
[43] Telegraph, March 4, 1902
[44] Telegraph, March 3, 1902
[45] Telegraph, March 13, 1902
[46] Telegraph, March 24, 1902
[47] Star-Independent, March 26, 1902
[48] Telegraph, March 28, 1902
[49] The Pennsy, July 1953, p.10
[50] 125th Celebration Marysville, Pa., p. 29
[51] Patriot, Telegraph, Star-Independent, March 31, 1902
[52] Final cost listed on PRR Plan 7034
[53] Reported in Locomotive Engineers' Journal, September 1952

Chapter 6

[1] Patriot, April 2, 1902
[2] Telegraph, April 10, 1902
[3] Telegraph, April 19, 1902
[4] Star-Independent, April 21, 1902
[5] Telegraph, April 18, 1902
[6] Telegraph, May 30, 1902
[7] Telegraph, May 13, 1902
[8] Star-Independent, June 13, 1902
[9] Star-Independent, July 5, 1902
[10] Telegraph, October 2, 1902
[11] Star-Independent, June 16, 1902
[12] Telegraph, January 2, 1903
[13] Telegraph, July 25, 1902
[14] Telegraph, October 12, 1901
[15] Telegraph, October 9, 15, 1902
[16] Telegraph, August 15, 1902
[17] Telegraph, October 13, 1902
[18] Telegraph, December 27, 1902
[19] Telegraph, March 23, 1903
[20] Telegraph, November 21, 1902
[21] Telegraph, February 9, 1903
[22] Telegraph, April 4, 1903
[23] Correspondence, Brown to Nauman, December 8, 1902, and Nauman to Brown, December 9, 1902, Hagley
[24] Correspondence, Nauman to McCaleb, December 12, 1902, Hagley
[25] Correspondence, Nauman to McCaleb, December 12, 1902, Hagley
[26] Altoona Tribune, January 26, 1903
[27] Telegraph, January 23, 1903
[28] Patriot, October 22, 1903
[29] Star-Independent, October 22, 1903
[30] Mechanicsburg Daily Journal, January 28, 1904
[31] Mechanicsburg Daily Journal, December 7, 1904
[32] Coatesville Times, September 24, 1904
[33] All dates and figures listed here from The Pennsylvania Railroad Company Description and History of Important Bridges and Stations, 1945

Chapter 7

[1] Schotter, p. 436
[2] Dubin, Arthur, Some Classic Trains, Milwaukee, Wis., Kalmbach Publishing Co., 1960
[3] Middleton, William D., Manhattan Gateway, Waukesha, Wis., Kalmbach Publishing Co., 1996, p. 70
[4] B&K, p. 525
[5] PRR Information, June 1928, p.1
[6] PRR Form 43 timetable, November 24, 1912
[7] Telegraph, May 30 and June 1, 1903
[8] 125th Celebration Marysville book, p. 31
[9] Telegraph, November 1, 1913
[10] Telegraph, January 8, 1914
[11] C&C, Vol. 1, p. 80

[12] Brignano, Mary, & McCullough, Search for Safety, published by Union Switch & Signal, 1981, p. 142
[13] Schotter, pp. 365, 374
[14] Schotter, p. 355
[15] Wornom, Douglas, History Passenger Car and Through Train Service 1849-1947, Chicago, Owen Davies, 1974, pp. 16-17
[16] Telegraph, March 27, 1916
[17] Pennsylvania Highways 1927, p. 10
[18] 125th Celebration, Marysville, Pa., p. 32
[19] Cupper, Crossroads, p. 80
[20] Cupper, Crossroads, p. 59
[21] Cupper, Crossroads, p. 81
[22] From West to East in Style, TRAINS, April 1984, p. 52

Chapter 8
[1] PRR Annual Report for 1932, p. 48
[2] Reutter, Mark, Railroad History Special, Diesel Revolution, p. 46
[3] Evening News, March 5, 1936
[4] PRR History of the Floods of 1936 and 1937, p. 59
[5] PRR History of the Floods of 1936 and 1937, p. 80
[6] PRR History of the Floods of 1936 and 1937, p. 80
[7] PRR Form 1 timetable, June 15, 1938
[8] Patriot, January 1, 1934, July 22, 1938
[9] Patriot, July 26, 1939
[10] Patriot, August 30, 1939
[11] Authorization for expenditure dated July 29, 1941, Hagley
[12] Request for approval dated June 8, 1942, Hagley
[13] Dominic Pinci, interview with author, March 22, 2002
[14] New York Times, June 28, 1942
[15] Taber, Encyclopedia and Atlas, p. 459
[16] Reutter, Railroad History 170, p. 26
[17] Cupper, Crossroads, p. 128
[18] PRR Annual Report for 1947, p. 37
[19] PRR Timetable No. 8, April 28, 1946, General Order 101, p. 116
[20] Interview by author with Susquehanna Appalachian Trail Club co-founder Ralph Kinter, March 21, 2002
[21] Modern Railroads, November 1956, p. 14
[22] Modern Railroads, November 1956, p. 74
[23] The Pennsy, September-October 1957, p. 1

Chapter 9
[1] PRR Philadelphia Division Timetable No. 14, April 24, 1949, pp. 94-95. Steam engines restricted were classes T1, HH1, S, and Q.
[2] PRR 114th annual report for 1960, PRR 116th annual report for 1962
[3] Harrisburg Evening News, February 9, 1961
[4] Amtrak system timetable, May 1, 1971
[5] Penn Central Form 16 timetable, December 1970
[6] The National Weather Service considers a rainstorm to be a hurricane if winds exceed 39 mph; otherwise NWS terms it a tropical storm.
[7] Penn Central Post, August-September 1972, p. 1
[8] National Oceanic and Atmospheric Administration press release 97-R228 dated June 11, 1997
[9] Penn Central Post, August-September 1972, p. 1
[10] Penn Central Post, October-November 1972, p. 1
[11] Penn Central Post, September 1973, p. 7

Chapter 10
[1] Harrisburg Patriot, August 29, 1977
[2] Harrisburg Patriot, August 31, 1977
[3] Interview with Abram Burnett, former supervisor of operating rules, February 24, 2002.
[4] Department of Environmental Protection news release, August 22, 1997
[5] Department of Environmental Protection news release, August 29, 1997
[6] Department of Environmental Protection news release, September 19, 1997
[7] NS Operations Bulletin 216, November 9, 2001

Selected Bibliography

Books

______, 125th Celebration, Marysville, Pa., 1991

_____, Battles and leaders of the Civil War; being for the most part contributions by Union and Confederate officers. Based upon "The Century War Series." Edited by Robert Underwood Johnson and Clarence Clough Buel, of the editorial staff of "The Century Magazine." Publisher: New York, Century Co. [c1884-1889]

_____, Guide for the Pennsylvania Railroad, Including the Entire Route with all Its Windings, Objects of Interest, and Information Useful to the Traveller, Philadelphia—T. K. and P. G. Collins, Printers, 1855

_____, History of that part of the Susquehanna and Juniata Valleys embraced in the Counties of Mifflin, Juniata, Perry, Union and Snyder, in the Commonwealth of Pennsylvania: Philadelphia, Everts, Peck & Richards, 1886

Burgess, George, and Miles C. Kennedy, Centennial History of the Pennsylvania Railroad Co., Philadelphia, published by the railroad, 1949

Coverdale & Colpitts, consulting engineers, The Pennsylvania Railroad Company Corporate, Financial and Construction History of Lines Owned, Operated and Controlled to December 31, 1945, New York, 1945

Cupper, Dan, Crossroads of Commerce, The Pennsylvania Railroad Calendar Art of Grif Teller, Richmond, Vt., Great Eastern Publishing Co., 1992

Gunnarsson, Robert L., The Story of the Northern Central Railway, Sykesville, Md., Greenberg Publishing Co., 1991

Haupt, Herman, Haupt on Bridge Engineering, New York, D. Appleton & Co., 1853
_____, Reminiscences of General Herman Haupt, self-published, 1901

Miller, William J., The Training of an Army: Camp Curtin and the North's Civil War (Also republished as Civil War City: Harrisburg, Pennsylvania 1861-1865), Shippensburg, Pa., White Mane Publishing Co., 1990

Morgan, Ed, Quarries of Curwensville, the People the Legacy, undated

Sipes, William B., The Pennsylvania Railroad, its origin, construction, condition, and connections, PRR Passenger Department, 1875

Schotter, Howard, Growth and Development of the Pennsylvania Railroad, Philadelphia, published by the railroad, 1927

Taber, Tom, Railroads of Pennsylvania Encyclopedia and Atlas, Muncy, Pa., published by the author, 1987

Tyrrell, Henry G., History of Bridge Engineering, the author, Chicago, 1911

Wilson, William Bender, History of the Pennsylvania Railroad, published by the railroad, Philadelphia, 1895

Annual Reports

Northern Central Railway
Pennsylvania Railroad Co.

Manuscript

Coverdale & Colpitts, The Pennsylvania Railroad Company, Description and History of Important Bridges and Stations, 1945

Periodicals

Altoona Tribune, Altoona
Democratic Union, Harrisburg
Harrisburg Telegraph, Harrisburg
The Patriot, Harrisburg
Patriot-News, Harrisburg
The Star-Independent, Harrisburg
Journal of Lancaster County Historical Society
Penn Central Post
Modern Railroads
Railroad Gazette
Railroad History
The Pennsy
Trains Magazine

Index

The Susquehanna can freeze over, sometimes for months at a time. Rodney DiPaolo

Acknowledgements

Thanks are due to many people who offered support or shared their collections, recollections, or expertise, to make this book more complete and more accurate. Kurt Bell, curator of the Railroad Museum of Pennsylvania, guided me to important holdings there. Thomas E. Boothby and Mark Brown at the architectural engineering department at Penn State University shared their knowledge of stone-arch bridge history. Abe Burnett of New Cumberland, Pa., a veteran railroader, patiently answered phone calls and e-mails. Eric DeLony, chief of the Historic American Engineering Record, shared his insights over a Thai dinner in Lancaster, Pa. Eileen Drelick provided guidance in her role as a representative of American Premier Underwriters, the former Penn Central Co. Rich Hill and the staff of the State Library newspaper room were, as ever, helpful. Valerie Miller at The Library Company of Philadelphia assisted in identifying William H. Rau photographs of the bridge. Spencer Nauman Jr. and his cousin, Ellen Martin of Haverford, Pa., shared the history of the Nauman family. Dominic Pinci of Harrisburg and his brother Paul Pinci of Marysville shared World War II recollections; Col. Thomas Sweeney at the Army War College in Carlisle, Pa., reviewed and commented on the sections dealing with military operations. Bookseller Mary Ellen Tarman of Hummelstown, Pa., offered support on practical aspects of book production and marketing. Several former Conrail bridge engineers took time to describe the 1997 spandrel wall failure and cleanup. Others who provided support were Bob Troup of Duncannon, Pa., a member of both the Rockville Bridge U.S. Stamp Committee, and the Harrisburg Chapter of the National Railway Historical Society. Others who were helpful were Bill Hummel of Marysville National Bank, historian William D. Middleton, and Artie Tafoya of the Appalachian Brewing Co.

Many collectors and photographers were generous in sharing their material, including Fred Abendschein of Columbia, Pa., Jim Bradley of Harrisburg, Mark Brinck of Harrisburg; Frank Rupley Daum of Marysville, Duane Hammaker of Ann-Katherine's Antiques in Duncannon; Kerry L. Jury of Mechanicsburg, Pa., Bennett Levin of Washington Crossing, Pa.; Robert R. Malinoski of Camp Hill, Pa.; Walter Minnich Jr., of Quarryville, Pa.; Ken Murry of Mountville, Pa.; Dr. Jeremy Plant and his son Brian Plant of Hummelstown; brothers Walt and Eric Ready of Marysville; Jim Roberts of Marysville; John and Peggy Smith of Peggy's Antiques in Duncannon; John W. Smith of Camp Hill, Pa.; Fred Wertz of Mechanicsburg, Pa.; Don Wood of Matawan, N.J.; Ted Xaras of Lansdowne, Pa.; and Marty Zak of Collegeville, Pa.

My longtime friend Rich Roberts agreed to serve as copy editor, a usually thankless task. Another longtime business colleague and friend, Paul K. Withers, served not only as publisher but also as layout designer. Thanks also go to the staff of Paulhamus Litho.

Finally, Rodney DiPaolo, a Rockville Bridge historian and collector who with his family lives along the former Pennsylvania Railroad main line at Cove, west of Marysville, was a constant source of encouragement, ideas, contacts, memorabilia, and photos. It was Rodney's canoe and his familiarity with the Susquehanna that enabled me to appreciate the full expanse of the bridge for the first time.

Dan Cupper

Harrisburg, Pa., March 24, 2002